SHORTCUTS: BOOK 1
STRUCTURE AND FABRIC

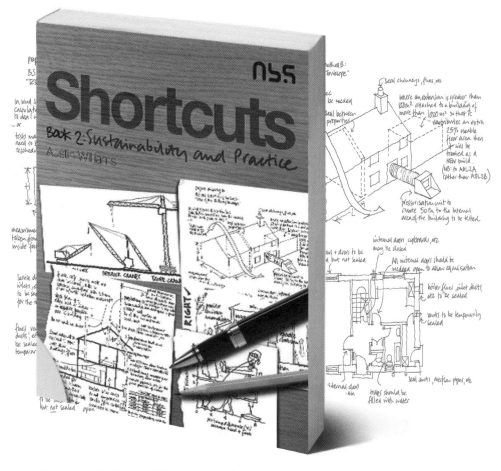

Customer Returns Form

Please address all returns to:
RIBA Bookshops Returns Department, 28-42 Banner Street, London EC1Y 8QE, UK
See also Terms and Conditions (item 3) overleaf

Company name _____

Address _____

_____

_____

Contact _____ **Telephone number** _____

Contact email address _____

Account number _____ Invoice number _____

ISBN	Item returned	Qty	Reason (insert code A-E, see below)

Reason for returning items:

(a) Wrong item ordered
(b) Damaged (specify in comments below)
(c) Duplicated item

(d) Goods not as invoice
(e) Other (specify in comments below)

Refund/Credit requested (refund only applicable if original payment made by cheque or credit card):

❏ **Refund** ❏ **Credit**

Comments

_____

_____

_____

_____

Please note: we are unable to refund costs of return postage where items were ordered incorrectly, or were not required.

❏ I have read and accept the Terms and Conditions overleaf (please tick)

RIBA Bookshops Mail Order
T +44 (0)20 7256 7222   F +44 (0)20 7374 2737
sales@ribabookshops.com   www.ribabookshops.com
Opening hours: Mon–Fri 09.00–17.30

RIBA ☷ **Bookshops**

Terms and Conditions

1. Payment
- Cheque/International Money Order, which must be in Sterling and drawn on a UK bank (Payable to RIBA Enterprises)
- RIBA Bookshops Credit Account – applications are welcome
- Credit/Debit Card: MASTERCARD/VISA/SWITCH/MAESTRO/DELTA/EUROCARD/SOLO/AMERICAN EXPRESS/ELECTRON
- Payment by credit/debit card is taken at the point the order is placed online or when the order is processed by our Customer Service team (telephone and fax orders). Where goods are out of stock or are forthcoming, payment will be taken and goods will be dispatched automatically when available. We will advise customers of any problems with their order. Customers awaiting outstanding orders may cancel at any time and receive a full refund (see returns and refunds).

2. Delivery and Despatch
- **Within the UK:** Goods weighing less than 2kg are sent by First Class post, while those weighing more than 2kg are sent by 1–2 day courier service
- Most UK orders are delivered within 3 working days
- A signature may be required on delivery for orders weighing over 2 kg
- **Overseas:** Goods sent outside the UK are sent by standard courier services unless otherwise requested, taking approximately 7–21 working days
- Orders are normally despatched the following working day subject to availability
- Next working day delivery (UK mail order/telephone orders): normal post & packing charges apply plus a Priority Despatch Fee of £3. Orders must be received by 14.00 GMT Monday–Friday for delivery the next working day before 17.30 GMT.
- Next working day delivery (UK online orders): guaranteed next working day delivery for orders £60 and under – normal post & packing charges apply plus Priority Despatch Fee of £3. Orders over £60 (with free postage entitlement) – £5 Priority Despatch Fee. Orders must be received by 14.00 GMT Monday–Friday for delivery the next working day before 17.30 GMT.
- Overseas orders: urgent delivery services are available at cost plus 20% administration fee. Please contact RIBA Bookshops on +44 (0)20 7256 7222 or email sales@ribabookshops.com for information
- The following rates are correct at the time of going to press, but are liable to change without prior notice

Postage & Packing UK
- Minimum postage charge is £3 for orders under £30
- Up to £100 in value, add 10% of the total order value (minimum charges apply)
- £100 – £200, add £10 flat fee
- Over £200, FREE postage
- Free postage for all UK orders placed online over £60 (calculated on the sub-total excluding VAT. Excludes downloadable products)

Postage & Packing Overseas
- Minimum postage charge is £10 for orders under £50
- Standard despatch charge add 20% to the total order value (minimum charges apply)

3. Returns and Refunds
- Any claim for non or late delivery of goods must be made within 28 days of placing the order
- A Returns Form (sent with your order) should accompany any returned items. The form is also available in the Help/Information section at www.ribabookshops.com, or can be emailed or faxed on request

- Customers who qualify for any free despatch offers will be asked to reimburse the postage costs in full, if any subsequent adjustment to the order results in the final order value being less than the qualifying amount
- Software products which have been unsealed are returned at the discretion of RIBA Bookshops
- No refunds can be issued for electronic documentation/downloads including RIBA Agreements Online and CIC Contracts Online
- Goods remain the property of RIBA Enterprises Ltd until paid for in full
- Please address all returns to:
  RIBA Bookshops Returns Department
  28–42 Banner Street,
  London EC1Y 8QE, UK

Return of Goods
- Customers wishing to return their order may do so up to 14 days from the invoice date for UK customers and 28 days for overseas customers
- A returns form should accompany any returned goods (see above)
- Goods must be returned in a re-saleable condition and customers must arrange for their return at their own expense
- A refund to the value of the goods (excluding despatch costs) will be made within 30 days

Damaged or Incorrect Goods
- Any damaged or incorrectly despatched goods should be returned with original packaging if possible, and a returns form (see above), within 14 days of the invoice date for UK customers and 28 days for overseas customers
- Where goods are damaged or incorrectly despatched, a complete refund for the goods and the despatch costs will be provided. RIBA Bookshops will also bear the full cost of returning the goods, either by refund, or arranging to have the goods collected at our cost

4. Account Queries
- Any account queries should be raised with the Finance Department on
  Tel +44 (0)191 244 5510
  Email finance@ribaenterprises.com

RIBA Enterprises
The Old Post Office, St. Nicholas Street, Newcastle upon Tyne NE1 1RE, UK

5. Prices and Editions
- RIBA Bookshops believe all prices and details to be correct at the time of going to press, but these may be subject to alteration without prior notice

6. VAT
- At the time of going to press, books are zero rate, while most forms and contracts are subject to VAT at the standard rate. VAT is calculated on the net value of the goods, including the despatch costs for VATable goods, at the full VAT rate.

7. Trade Terms
- Available on application to:
  RIBA Bookshops
  15 Bonhill Street, London EC2P 2EA, UK

Where to find RIBA Bookshops

Mail Order
15 Bonhill Street, London EC2P 2EA, UK
Tel +44 (0) 20 7256 7222
Fax +44 (0) 20 7374 2737
Email sales@ribabookshops.com
Web www.ribabookshops.com

London, Central
RIBA
66 Portland Place, London W1B 1AD, UK
Tel +44 (0) 20 7256 7222
Fax +44 (0) 20 7637 7185
Email sales@ribabookshops.com

London, Chelsea
Ground Floor, North Dome
Design Centre Chelsea Harbour
Lots Road, London SW10 0XF, UK
Tel +44 (0) 20 7351 6854
Email chdc.bookshop@ribabookshops.com

Manchester
CUBE
113–115 Portland Street,
Manchester M1 6DW, UK
Tel + 44 (0) 161 236 7691
Fax + 44 (0) 161 236 1153
Email manchester@ribabookshops.com

Birmingham
Birmingham and Midland Institute
Margaret Street, Birmingham B3 3SP, UK
Tel +44 (0) 121 233 2321
Email birmingham@ribabookshops.com

Belfast
RSUA
2 Mount Charles, Belfast BT7 1NZ
Tel +44 (0) 2890 323 760
Email belfast@ribabookshops.com

Liverpool
RENEW Rooms
82 Wood Street, Liverpool L1 4DQ, UK
Tel +44 (0) 151 707 4380
Email liverpool@ribabookshops.com

Visit our website at **www.ribabookshops.com** for our full terms and conditions.

RIBA Bookshops

SHORTCUTS: BOOK 1
STRUCTURE AND FABRIC

AUSTIN WILLIAMS

nbs

© **RIBA Enterprises Ltd. 2008**

Published by NBS, part of RIBA Enterprises Ltd.
RIBA Enterprises Ltd., 15 Bonhill Street, London EC2P 2EA

ISBN 978 1 85946 321 5

Stock code 67742

British Library Cataloguing in Publications Data
A catalogue record for this book is available from the British Library.

Publisher: Steven Cross
Commissioning Editor: James Thompson
Project Editor: Anna Walters
Designed and typeset by Kneath Associates
Printed and bound by Latimer Trend, Plymouth

www.ribaenterprises.com
www.thenbs.com

PREFACE

Welcome to Shortcuts.

Shortcuts are at-a-glance guides designed to help construction professionals navigate the minefield of regulations, new technologies and the diverse (and sometimes conflicting) range of technical guidance documents.

This is the first book of Shortcuts, and it deals with matters of structure and fabric. It contains information on tree roots, mortar mixes, sulfate attack, soakaways, basements, acoustic loops, drainage, domestic lighting, rainscreen cladding, loft conversions and much more besides. Other topics range from wood warp to carpet weft, from DPCs to LEDs, and from soil nails to wood screws – from muck to brass, as they say.

Regulations governing construction are changing with increasing regularity these days. Whether it is creeping legislation or government consultation, it is becoming more and more difficult to keep up with the changes. However, with Shortcuts, I read the documents so that you don't have to. Each Shortcut provides a distillation of the necessary technical information, and a straightforward starting point for you to understand the important bits while not having to wade through the unnecessary waffle.

I hope that this book is useful. If you have any comments or suggestions on future topics, please don't hesitate to let us know.

Yours sincerely,

Austin Williams

CONTENTS

Part 1
STRUCTURE

1

ight in metres

lar (25)

European

Leyland
cypress (20)

hawthorn (10)

se chestnut (20) — pine (

sycamore (24)

wild cherry
(18)

(10)
laurel

holly (

(9.
magnolia

01: Lofty Aspirations
Utilising the roofspace

To most people the reality of loft living is a long way from Urban Splash's model of urban regeneration. A large proportion of planning applications simply relate to the conversion of a dusty attic space into an extra bedroom or two. It seems that top floor domestic extensions are more manageable investments than the chic warehouse alternatives.

Kate Barker's 'Review of Land Use Planning', published in 2006, addressed the need to build substantial volumes of new housing by freeing up development potential. Her recommendations include not only relaxing controls on building on the greenbelt, but also relaxing planning powers more generally. The report said that 'planning resources should be able to focus more on the larger-scale applications, rather than the small-scale permissions which have little impact on the wider public interest. To achieve this, the principle of the Householder Development Consent Review (that permitted development rights for householders should be extended based on an 'impact' principle) should be rolled out to minor applications.'

These *'minor applications'* weren't spelled out. However, Barker explicitly recommended reducing planning powers over domestic and commercial microgeneration proposals, and her words have been interpreted as extending permitted development rights to make it easier to build on, convert and extend domestic roofs. The government has generally accepted the recommendations in her report. As a result, ministers agreed that, from 1 October 2008, loft conversions are allowed without planning consent, *provided that* they meet certain conditions. For instance, they must not extend beyond the plane of the existing roof slope on the principal elevation fronting the highway.
(Note: Work to properties more than 20 m from a highway are permitted anyway, provided that they also satisfy all other criteria.) Essentially, this implies that a traditional rooflight with raised jambs for flashing, taking it above the plane of the roof, will not be permitted. But bear in mind that planning departments have discretionary powers, and decisions may vary between authorities.

Planning departments have discretionary powers and decisions may alter between authorities.

Other criteria that have to be satisfied before a loft may forego planning consent are:

■ The new volume must not exceed an additional 40 m³, for terraced houses (internal measurements)

■ The new volume must not exceed an additional 50 m³, for all other dwellings.

■ No extension may extend above the highest part of the roof.

■ The materials must be similar in appearance to the existing house.

■ There should be no verandas, balconies or raised platforms.

■ Side-facing windows must be obscure-glazed and any opening portion must be at least 1.7 m above the internal floor level.

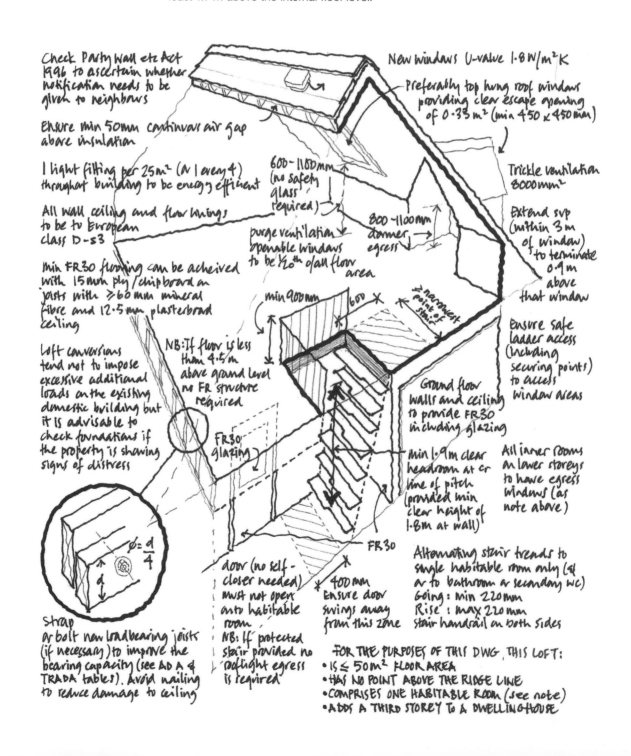

Check Party Wall etc Act 1996 to ascertain whether notification needs to be given to neighbours

Ensure min 50mm continuous air gap above insulation

1 light fitting per 25m² (N 1 every 4) throughout building to be energy efficient

All wall ceiling and floor linings to be to European class D-s3

Min FR30 flooring can be acheived with 15mm ply/chipboard on joists with >60 mm mineral fibre and 12.5mm plasterboard ceiling

Loft conversions tend not to impose excessive additional loads on the existing domestic building but it is advisable to check foundations if the property is showing signs of distress

Strap or bolt new loadbearing joists (if necessary) to improve the bearing capacity (see AD A & TRADA tables). Avoid nailing to reduce damage to ceiling

$\phi = \frac{d}{4}$

600-1100mm (no safety glass required)

purge ventilation openable windows to be 1/20th of all floor area

NB: If floor is less than 4.5m above ground level no FR structure required

FR30 glazing

min 900mm

650

800-1100mm dormer egress

narrowest point of stair

door (no self-closer needed) must not open onto habitable room NB: If protected stair provided no rooflight egress is required

400mm Ensure door swings away from this zone

FR30

New windows U-value 1.8W/m²K

Preferably top hung roof windows providing clear escape opening of 0.33 m² (min 450 x 450 mm)

Trickle ventilation 8000mm²

Extend svp (within 3 m of window) to terminate 0.9 m above that window

Ensure safe ladder access (including securing points) to access window areas

Ground floor walls and ceiling to provide FR30 including glazing

Min 1.9m clear headroom at cr line of pitch (provided min clear height of 1.8m at wall)

All inner rooms on lower storeys to have egress windows (as note above)

Alternating stair treads to single habitable room only (& or to bathroom or secondary wc) Going: min 220mm Rise: max 220 mm stair handrail on both sides

FOR THE PURPOSES OF THIS DWG, THIS LOFT:
• IS ≤ 50m² FLOOR AREA
• HAS NO POINT ABOVE THE RIDGE LINE
• COMPRISES ONE HABITABLE ROOM (see note)
• ADDS A THIRD STOREY TO A DWELLINGHOUSE

- There should be no roof extensions in designated areas (the Broads, national parks, areas of outstanding natural beauty, conservation areas and world heritage sites).

- Roof extensions, apart from those rising from hip to gable, must be at least 20 cm from the eaves.

Recent surveys indicate that more and more people are considering extending their family home rather than suffering the pain and expense of moving to another property. Ex housing minister Caroline Flint noted that, as the economy slows, these new planning rules will 'make a real difference to already-stretched family finances, making home improvement an increasingly attractive option.' Given the recent collapse of credit and the fall in the housing market, people seem not only to be moving less, but also building less.

Popular TV 'make-over' shows undoubtedly give the impression that extensions and conversions are easier to carry out than they actually are, but the financial equation is worth considering before selling up. At the time of writing, a typical four-bedroom house costs about £35,000 more than a three-bedroom property. Add to this, say, £7500 stamp duty plus removal costs and, by moving, you may be looking at gaining an extra bedroom and, perhaps, an extra bathroom for a total additional outlay of around £44,000. Compare that with the cost of converting a loft. Provided that the conversion is not structurally or architecturally complicated, a new attic bedroom and bathroom is likely to cost about £25,000, but the new facilities could add the aforementioned £35,000 to the existing property value.

The planning procedures for small-scale developments should become less onerous, implying that homeowners could bypass local authority planning departments for schemes comprising a loft extension (or a conservatory). The decision to liberalise permitted development for loft extensions omits the original recommendations of the Barker Review that only those extensions/conversions which have a 'non-marginal third-party impact' will evade planning consents. Under the new legislation, no such consideration need be given. However, it is still advisable that one gains consent from one's neighbours – and of others in the vicinity who may be affected by the work – prior to proceeding, since there still remains the possibility of them complaining.

Currently, if the loft conversion is a permitted development, the complaint will come to nought, as the application is dealt with on purely legal (rather than aesthetic) grounds. This has often caused disputes between neighbours, and Barker's more open forum of neighbour notification is intended to overcome the perceived underhandedness of the current system.

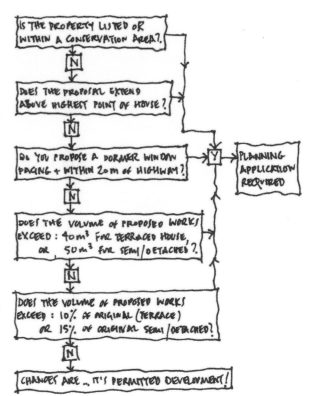

Even though there may be some people who will prefer to submit applications in the traditional way, the government estimates that the proposals will reduce 'minor' planning applications by around 80,000 per year. In 2007, around 340,000 domestic planning applications were submitted (approximately 50 per cent of all applications) of which about 90 per cent were approved.

Even without the new legislation, loft conversions have only rarely required planning permission. For example, new-build work to a detached or semi-detached property would have had to have exceeded 50 m³ or 15 per cent of the original volume, whichever is greater (the corresponding values for a terraced property are 40 m³ or 10 per cent of the original house volume).

Note: Permitted development usually does not apply if the work affects a listed building, is in a conservation area, contravenes existing planning constraints, or is subject to local restrictions. Therefore, it is advisable to consult with the local authority at the earliest opportunity. But as the Department of Communities and Local Government admits, 'It may not always be clear whether a planning application is needed – even to planning officers.'

Should a complaint be made, the local authority planning inspector may visit to assuage concerns and, given the ten criteria cited above, this should be straightforward in legitimate cases. In more complicated cases, a 'Lawful Development Certificate' (LDC) will be required, certifying that

> *Even though there may be some people who will prefer to submit applications in the traditional way, the government estimates that the proposals will reduce 'minor' planning applications by around 80,000 per year.*

the works do not require full plans submission. An LDC is no more than a statement of legal fact that guarantees the lawfulness of the work it certifies. An application for an LDC should detail (with supporting documentation) all relevant issues, such as existing planning permissions, enforcement notices or certificates that cover the proposed use or development, as well as an exact description of the proposed use. The onus is on the applicant to show that planning permission is not required. As part of the current review, LDCs may be deregulated to approved third parties. Until then, the current system prevails and, in theory, an LDC takes around 4–8 weeks to issue, compared with a full plans approval which may take several months to assess. An LDC has no expiry date, unlike formal planning permission which typically expires within 3–5 years.

Until April 2007, when Approved Document B (AD B) 'Fire Safety' came into force, if a design incorporated an open-plan loft access, then all doors to habitable rooms off the stairwell had to be provided with self-closers. This is no longer the case, but the government is urging homeowners to keep closed all doors that protect the escape route, especially at night.

Under the new Approved Document B provisions, a two-storey house with a loft conversion is treated as a three-storey property, requiring escape via a protected stair. Inner rooms are not permitted in the converted loft area, and inner rooms on lower storeys must have (or be provided with) egress windows, which may be lockable to protect children provided that they have a safe release mechanism. Existing doors accessing the existing staircase must be replaced with fire-resisting doors to ensure that there is a fire-protected route to the final ground-floor exit; the exception is doors to non-habitable rooms such as bathrooms and cloakrooms. If it is considered undesirable to replace existing doors, e.g. if they are of historical or architectural merit, it may be possible to retain them or upgrade them to an acceptable standard. Internal glazing along protected escape routes within fire-resisting structures must also be fire-resisting. Self-closing devices are not required on fire-resisting doors.

New rooms in a domestic loft space should be design/risk assessed in accordance with CDM 2007: Regulation 11. In lofts where the floor is less than 4.5 m above ground level and where it is not feasible to provide a protected stair, the loft floor should have 30 minutes fire resistance (FR30) and have a rooflight or similar to provide secondary escape. Windows or rooflights for emergency egress from a loft room (or from a second-floor room where the loft creates a third storey) should meet the following criteria:

- Width and height: not less than 450 mm

- Clear openable area: not less than 0.33 m²

- Bottom of openable area: not more than 1100 mm above finished floor level

The window or rooflight should enable the person escaping to reach a place free from danger from fire and, especially where the floor is greater than 4.5 m above external ground level, should be in a part of the roof that can be reached by somebody providing outside assistance using a ladder.

RELEVANT PARTS OF THE GENERAL PERMITTED DEVELOPMENT ORDER

Part 1 gives permission for the construction of extensions, porches, free-standing buildings or swimming pools, external hard surfacing, containers for oil, satellite antennae and alterations to a roof. It sets height and volume limits for new developments.

Part 2 allows the erection of boundary enclosures, including walls, fences and gates subject to height restrictions, exterior painting and the provision of hardstandings.

References

Architects' Journal (2001) *'Dampener on loft living'*, AJ 05.04.2001.

Barker, K. (2006) *'Barker Review of Land Use Planning: Final Report – Recommendations'*, HMSO.

BS EN 14975 (2006) *'Loft ladders – Requirements, marking and testing'*, BSI.

Coutts, J. (2006) *'Loft Conversions'*, Blackwell.

Williamson, L. (2000) *'Loft Conversions: Planning, Managing and Completing Your Conversion'*, The Crowood Press.

RECOMMENDED READINGS
Building Research Establishment (2006) Good Building Guide 69: Part 1 *'Loft conversion: structural considerations'*, BRE.

Bulding Research Establishment (2006) Good Building Guide 69: Part 2 *'Loft conversion: safety, insulation and services'*, BRE.

Department of Communities and Local Government (2006) *'Householder Development Consents Review – Steering Group Report'*, DCLG.

Jones, E, & Sparks, L. (2006) *'The Householder Development Consents Review: Appendix 1, Making the System More Proportionate'*, DCLG.

Department of Communities and Local Government (2008), *'The Town and Country Planning (General Permitted Development) (Amendment) (England) Order 2008'*, TSO.

TRADA Technology (1996) *'Loft conversion guide for contractors and designers'*, TRADA.

TRADA Technology (2003) Wood Information Sheet 0/12 *'Room in the roof construction for new houses'*, TRADA.

02: Timber Spans
Rafters, floor and ceiling joists

The span-load tables used to be printed in Approved Document A: (AD A) Structure (1994 edition) but were removed from the current (2004) version. Fortunately, equivalent data is included within BS 8103-3 (1996) 'Structural design of low-rise buildings', and the AD A: 1994 is still handy. We have redrafted a few read-off tables for some common structural timber situations.

This Shortcut is intended simply to provide an at-a-glance guide to indicate the possible length of span of some standard structural timber sections under normal loading conditions. This Shortcut, perhaps more so than any of the others, should be seen as an initial aid to design only. Even though these tables were good enough for architects over the course of the 1990s, it is always advisable – where structural stability is concerned – to consult an engineer to give you the final nod.

BS 8103-3 identifies a wide range of dos and don'ts for the satisfactory use of structural timber, many of them seemingly ignored (or unknown) in the domestic construction sector. For instance:

- Using the span tables to size bathroom joists is only acceptable if the joists under the bath supports are doubled up.

- Strutting is necessary where the span of a floor exceeds 2.5 m, but herringbone strutting should not be used where the distance between joists is more than 'approximately three times the depth of the joist', e.g. possibly 195 × 38 mm joists at 600 mm. In these instances, solid struts at least three-quarters of the joist depth should be used.

These two nuggets were originally written into the 1994 edition of Approved Document A (AD A) but do not appear in the latest edition. The full span tables are published in TRADA Technology Design Aid DA 1/2004, and similar information is contained in BS 8103-3. The key difference between the original AD A and the contemporary information is simply that the strength class designations SC3/SC4 have been replaced by European strength classes. Fortunately, the new strength classes C16 and C24 are, to all intents and purposes, equivalent to the SC3 and SC4 classes of AD A: 1994.

When reading off the tables, note that the bearing length of a joist is assumed to be 38 mm for C16 timber, and 45 mm for C24 timber. For situations where end wane (non-square edge) is not permitted, the bearings are 20 mm and 24 mm respectively. However, BS 8103 states that, 'where a joist bears directly onto a plate on masonry, the end bearing 'should be at least 90 mm to provide restraint to the masonry wall'.

Maximum clear span of ceiling joists (metres) for timber of **Strength Class C16**

Size of joist (mm x mm)	Dead load (kN/m²) excluding the self-weight of the joist					
	Not more than 0.25			More than 0.25 but not more than 0.5		
	Spacing of joists (mm)					
	400	450	600	400	450	600
	Maximum clear span of joist (m)					
38 x 72	1.145	1.135	1.110	1.110	1.095	1.055
38 x 97	1.735	1.715	1.665	1.665	1.640	1.575
38 x 122	2.365	2.335	2.250	2.250	2.210	2.110
38 x 147	3.015	2.970	2.850	2.850	2.795	2.655
47 x 72	1.270	1.260	1.225	1.225	1.210	1.170
47 x 97	1.915	1.895	1.835	1.835	1.805	1.730
47 x 122	2.600	2.565	2.465	2.465	2.420	2.310
47 x 147	3.300	3.250	3.110	3.110	3.050	2.900

CEILING JOISTS:

Maximum clear span of ceiling joists (metres) for timber of **Strength Class C24**

Size of joist (mm x mm)	Dead load (kN/m²) excluding the self-weight of the joist					
	Not more than 0.25			More than 0.25 but not more than 0.5		
	Spacing of joists (mm)					
	400	450	600	400	450	600
	Maximum clear span of joist (m)					
38 x 72	1.210	1.200	1.170	1.170	1.160	1.120
38 x 97	1.835	1.815	1.755	1.755	1.730	1.660
38 x 122	2.495	2.460	2.370	2.370	2.940	2.790
38 x 147	3.175	3.125	2.995	2.995	2.940	2.790
47 x 72	1.345	1.330	1.300	1.300	1.280	1.240
47 x 97	2.025	2.000	1.930	1.930	1.900	1.825
47 x 122	2.740	2.700	2.595	2.595	2.550	2.430
47 x 147	3.470	3.415	3.270	3.270	3.810	3.605

FLOOR JOISTS:

Max clear span of floor joists (metres) for timber of **Strength Class C16** (supporting no loadbearing partitions)

Size of joist (mm x mm)	Dead load (kN/m²) excluding the self-weight of the joist								
	Not more than 0.25			More than 0.25 but not more than 0.5			More than 0.5 but not more than 1.25		
	Spacing of joists (mm)								
	400	450	600	400	450	600	400	450	600
	Maximum clear span of joist (m)								
38 x 72	1.130	1.010	0.765	1.060	0.950	0.730	0.920	0.830	0.650
38 x 97	1.825	1.690	1.300	1.715	1.555	1.210	1.420	1.300	1.035
38 x 122	2.480	2.385	1.925	2.365	2.215	1.755	1.945	1.790	1.445
38 x 147	2.980	2.865	2.510	2.850	2.705	2.330	2.450	2.290	2.270
38 x 170	3.440	3.305	2.870	3.280	3.095	2.690	2.805	2.645	2.265
38 x 195	3.935	3.750	3.260	3.720	3.515	3.055	3.185	3.010	2.605
47 x 97	2.020	1.910	1.580	1.920	1.815	1.460	1.665	1.525	1.225
47 x 122	2.660	2.560	2.300	2.545	2.450	2.085	2.225	2.080	1.695
47 x 147	3.200	3.075	3.785	3.060	2.945	2.605	2.720	2.570	2.170
47 x 170	3.690	3.550	3.185	3.530	3.395	2.985	3.115	2.940	2.550
47 x 195	4.220	4.060	3.615	4.040	3.890	3.390	3.535	3.340	2.900
63 x 97	2.315	2.195	1.915	2.190	2.080	1.822	1.930	1.840	1.530
63 x 122	2.930	2.820	2.565	2.805	2.700	2.450	2.525	2.425	2.085
63 x 147	3.515	3.385	3.080	3.370	3.240	2.950	3.035	2.920	2.575
63 x 170	4.055	3.905	3.555	3.885	3.740	3.400	3.500	3.370	2.950
63 x 195	4.630	4.465	4.070	4.440	4.275	3.895	4.005	3.850	3.350
63 x 220	5.060	4.920	4.580	4.905	4.770	4.365	4.510	4.300	3.750
75 x 147	3.720	3.580	3.265	3.560	3.430	3.125	3.215	3.090	2.805
75 x 170	4.280	4.125	3.765	4.105	3.955	3.605	3.705	3.570	3.210
75 x 195	4.830	4.700	4.305	4.680	4.520	4.125	4.240	4.080	3.645
75 x 220	5.265	5.125	4.790	5.110	4.970	4.640	4.740	4.595	4.070

NOTES:

1. Dimensions shown in the tables relate to:

■ Normal domestic accommodation of under three storeys and containing no more than four self-contained dwellings per floor

■ Single-storey garages or domestic extensions (with the above limitations)

■ Pitched roofs of 12 m maximum span, where the load <12 kN/m

■ Roof area <200 m²

■ Design wind speeds not exceeding 44 m/s

■ Wall height not exceeding 15 m in any location

■ Any edge of a pitched roof (with no parapets) not exceeding 10 m (AD A: 2004 recommends that in small, single-storey, non-residential buildings, the longest width or length must not exceed 9 m)

■ Non-trimming timbers.

2. Sizes for floor joists relate to BS EN 336 sizes and tolerances for sawn and processed timbers to class 1 and class 2 respectively, whereas the rafters make reference to class 1 only. It is advisable to check on the availability of timber sizes.

3. Rafter sizes relate to clear span dimensions, calculated in accordance with BS 5268-7.

FLOOR JOISTS:

Max clear span of floor joists (metres) for timber of **Strength Class C24** (supporting no loadbearing partitions)

Size of joist (mm x mm)	Dead load (kN/m²) excluding the self-weight of the joist								
	Not more than 0.25			More than 0.25 but not more than 0.5			More than 0.5 but not more than 1.25		
	Spacing of joists (mm)								
	400	450	600	400	450	600	400	450	600
	Maximum clear span of joist (m)								
38 x 72	1.260	1.185	1.020	1.215	1.145	0.989	1.105	1.045	0.870
38 x 97	1.935	1.830	1.585	1.835	1.740	1.510	1.635	1.550	1.355
38 x 122	2.580	2.480	2.200	2.465	2.395	2.075	2.180	2.070	1.825
38 x 147	3.100	2.980	2.710	2.970	2.850	2.585	2.665	2.560	2.305
38 x 170	3.580	3.440	3.125	3.425	3.290	2.990	3.080	2.955	2.675
38 x 195	4.095	3.940	3.580	3.920	3.770	3.420	3.525	3.385	3.065
47 x 97	2.140	2.025	1.760	2.025	1.920	1.675	1.795	1.705	1.500
47 x 122	2.770	2.660	2.420	2.650	2.550	2.290	2.380	2.265	2.005
47 x 147	3.325	3.200	2.910	3.185	3.060	2.780	2.865	2.750	2.495
47 x 170	3.835	3.690	3.360	3.670	3.535	3.210	3.305	3.180	2.880
47 x 195	4.385	4.220	3.845	4.200	4.045	3.675	3.785	3.640	3.300
63 x 97	2.430	2.320	2.030	2.305	2.190	1.925	2.030	1.930	1.710
63 x 122	3.045	2.930	2.670	2.915	2.805	2.550	2.625	2.525	2.270
63 x 147	3.665	3.520	3.205	3.500	3.370	3.065	3.155	3.035	2.760
63 x 170	4.210	4.055	3.700	4.035	3.885	3.540	3.640	3.505	3.185
63 x 195	4.770	4.635	4.230	4.610	4.445	4.050	4.165	4.010	3.645
63 x 220	5.200	5.060	4.730	5.045	4.905	4.560	4.675	4.510	4.105
75 x 147	3.860	3.720	3.390	3.700	3.565	3.250	3.340	3.215	2.925
75 x 170	4.445	4.285	3.910	4.265	4.110	3.745	3.855	3,71	3.375
75 x 195	4.965	4.830	4.470	4.815	4.685	4.285	4.405	4.245	3.860
75 x 220	5.415	5.270	4.930	5.250	5.110	4.775	4.875	4.740	4.350

RAFTERS:

Max clear span of common/ jack rafters (m) for timber of **Strength Class C16** (>22.5° but no more than 30°)

Size of rafter (mm x mm)	Dead load (kN/m²) notwithstanding the self-weight of the timber including **Imposed Load of 0.75 kN/m²**								
	Not more than 0.50			More than 0.5 but not more than 0.75			More than 0.75 but not more than 1.00		
	Spacing of joists (mm)								
	400	450	600	400	450	600	400	450	600
	Maximum clear span of joist (m)								
38 x 100	2.175	2.130	2.005	2.005	1.955	1.820	1.875	1.820	1.680
38 x 125	2.965	2.855	2.595	2.740	2.660	2.435	2.540	2.455	2.245
38 x 150	3.545	3.415	3.105	3.335	3.210	2.915	3.165	3.045	2.715
47 x 100	2.550	2.455	2.230	2.345	2.280	2.095	2.180	2.115	1.945
47 x 125	3.175	3.060	2.785	2.985	2.875	2.615	2.835	2.725	2.480
47 x 150	3.795	3.655	3.330	3.570	3.440	3.130	3.390	3.265	2.970

RAFTERS:

Max clear span of common/ jack rafters (m) for timber of **Strength Class C24** (>22.5° but no more than 30°)

Size of rafter (mm x mm)	Dead load (kN/m²) notwithstanding the self-weight of the timber including **Imposed Load of 0.75 kN/m²**								
	Not more than 0.50			More than 0.5 but not more than 0.75			More than 0.75 but not more than 1.00		
	Spacing of joists (mm)								
	400	450	600	400	450	600	400	450	600
	Maximum clear span of joist (m)								
38 x 100	2.475	2.380	2.165	2.325	2.235	2.030	2.205	2.120	1.925
38 x 125	3.080	2.965	2.700	2.895	2.790	2.530	2.750	2.645	2.400
38 x 150	3.685	3.550	3.230	3.465	3.335	3.035	3.290	3.165	2.875
47 x 100	2.650	2.550	2.320	2.490	2.400	2.180	2.365	2.275	2.065
47 x 125	3.300	3.180	2.895	3.105	2.990	2.720	2.945	2.835	2.580
47 x 150	3.940	3.780	3.460	3.710	3.575	3.255	3.525	3.395	3.085

sizes for floor joists relate to BS EN 336's sizes and tolerances for sawn and processed timbers to class 1 and class 2 respectively, whereas the rafters make reference to class 1 only.

Loading: 16 mm tongued and grooved softwood flooring should only be used on joists set at centres up to 505 mm; for joists at 600 mm centres, the flooring must be at least 19 mm. To span 450 mm, a minimum of 18 mm particleboard is required; 600 mm spans requires 22 mm thickness.

The minimum thickness of plywood flooring laid over (at least three) joists fixed at 600 mm centres is:

■ Finnish birch or conifer, grade I/II, sanded – 18 mm

■ American Construction and Industrial, grade C-D Exposure 1 or C-C Exterior, unsanded – 18 mm

■ Swedish softwood, P30 grade, unsanded – 16 mm

■ Canadian Douglas fir, Select tight face or Select grade, unsanded – 15.5 mm.

Wallplates: Timber wallplates should be at least 70 mm wide, with a minimum thickness of 38 mm (although in Scotland 25 mm thickness is often used). Notches, trims and holes are to be as shown in the drawing.

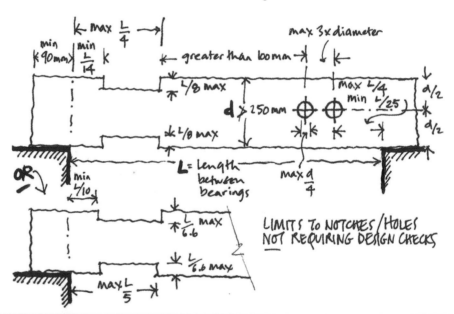

References

BS 5268 (1996) 'Structural use of timber: Code of practice for timber frame walls. Section 6.2: Dwellings not exceeding four storeys', BSI.

BS EN 1990 (2002) 'Eurocode 0: Basis of structural design', BSI (with National Annex [2004]).

BS EN 1991 (2002) 'Eurocode 1-1: Actions on structures: General actions. Densities, self-weight, imposed loads for buildings', BSI (with National Annex [2005]).

BS EN 1995-1-1 (2004) 'Eurocode 5: Design of timber structures. General. Common rules and rules for buildings', BSI (with National Annex [2006]).

Scottish Executive (2007) 'Technical Guidance. Section 1: Non-domestic: Structure', SE.

TRADA (2004) Technology Design Aid DA 1/2004 'Span tables for solid timber members in floors, ceilings and roofs (excluding trussed rafter roofs) for dwellings', TRADA Technology Ltd 2004 (Reprinted with 2005 and 2007 amendments).

RECOMMENDED READINGS

BS 5268 (2002) 'Structural use of timber – Part 2: Code of practice for permissible stress design, materials and workmanship', BSI.

BS 6399 (1996) 'Loading for buildings – Part 1: Code of practice for dead and imposed loads', BSI.

BS 6399 (1998) 'Loading for buildings – Part 3: Code of practice for imposed roof loads', BSI.

BS 8103 (1996) 'Structural design of low-rise buildings – Part 3: Code of practice for timber floors and roofs for housing', BSI.

Department for Environment Transport and the Regions (1994) 'The Building Regulations 1991: Approved Document A: Structure', The Stationery Office.

Office of the Deputy Prime Minister (2004) 'The Building Regulations 2000: Approved Document A: Structure', NBS.

Scottish Executive (2007) 'Technical Guidance. Section 1: Domestic: Structure', SE.

TRADA (2006) 'Timber frame housing: UK structural recommendations', 3rd edn, TRADA.

03: Root Causes
Building near trees requires extra deep foundations

Foundation design needs to take account of the water-absorbing capacity of nearby trees as well as the potential groundswell caused by trees that have been removed. This guide highlights the dangers to the structure that may be caused by the proximity of certain species.

Most trees in the UK have a significant radial root system, sometimes extending out to a distance of 1 to 1.5 times the height of the tree. Severing just one of a tree's major roots during careless excavation for construction or services can cause the loss of up to 20% of the root system, undermining its water-absorption and also leaving it unstable in high winds. Therefore, when laying service pipes, if it is possible to tunnel under the root system then it is advisable to do so – and running the pipes under the middle of the tree (on plan) if necessary.

In general, 80–90 per cent of all tree roots are to be found in the top 600 mm of soil, and almost 99 per cent of the tree's total root length occurs within the topmost 1m of soil (with some variations depending on soil porosity). The undoubted nuisance that fine root systems create for the development of specific sites has to be weighed up against the importance that they play in soil stabilisation on sloping ground (acting in a similar way to geotextile matting).

Only a few mature species, like oak, pine and fir have significant central tap roots (those main central roots from which the others spread) and, in most instances, even these only extend downwards about 2 metres. So it is the radial tree roots that extend outwards that are of primary concern here; these can influence soil conditions well beyond the circumference of the tree's leaf canopy. But with around £400 million-worth of tree-related insurance claims per annum in the UK, what precautions are needed when building near them?

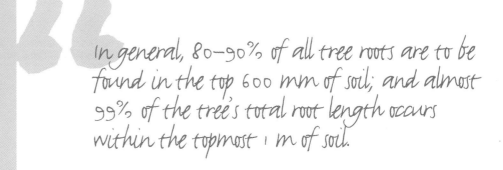

In general, 80–90% of all tree roots are to be found in the top 600 mm of soil; and almost 99% of the tree's total root length occurs within the topmost 1 m of soil.

It is always recommended that construction takes place as far away from trees and established vegetation as possible because the distance at which tree roots can detrimentally affect a building is quite significant. Actually, physical damage directly caused by proximate tree roots (e.g. roots exerting pressure on underground services, and cracking drains) is quite a rare occurrence; in fact, it is the secondary effects caused by the changing ground conditions near a tree – resulting from varying degrees of moisture removal – that should be the most important consideration for designers.

Check ground samples and liaise with the local authority and/or NHBC guidance to see whether they have any specific requirements to deal with local conditions. But, as a rule, for construction near trees, clay soils are more problematic than porous/sandy soils owing to their increased water retention and their potential to swell in heavy rain. Healthy trees on a site will take up large amounts of that water, removing it from the soil, which in some circumstances will be a help, but in other cases it can easily lead to clay soil shrinkage. Whether expansion or shrinkage, each action exerts a significant pressure on the foundations of nearby properties, potentially causing cracking and subsidence. It should be borne in mind that around 60 per cent of all the UK's housing is currently built on shrinkable clay.

It is important to remember though that removing trees will also affect ground conditions. A recently removed tree means that the moisture that would otherwise have been absorbed diurnally from the ground will remain, allowing the soil to swell and heave (the condition whereby water which would otherwise have been removed swells the clay soil causing pressures on trenches and slab foundations).

Similarly, deciduous trees have a seasonal impact on the ground moisture content, whereby winter rainfall rehydrates the dry summer soil with less of it being taken out of the ground by the dormant root action. Given that mature elms, oaks, horse chestnuts, planes and ashes can draw up to 50,000 litres of water a year from the surrounding soil, the consequent soil water retention, or frost, can lead to significant heave. Worst case examples have resulted in concrete slabs 'humping' as the soil expansion exerts an upward pressure on the floor slab. Similarly, trench foundations can crack with consequent movement affecting the structure above.

ZONES OF INFLUENCE

All trees have radial 'zones of influence' which diminish the further away from the tree the construction takes place. As a rule, it is recommended that properties be built at least a distance equivalent to the tree's height away from that tree. Attempts to insert a root barrier (e.g. a polypropylene or similar geomembrane) in order to dissuade root growth near a foundation, often encourage the roots to grow under and around it. Root barriers are also difficult to install near pre-existing trees, frequently compromising existing root growth with consequences as described above. However, while using root barriers to block an existing tree's root system will tend not to work, installing these barriers to control a *new* tree's growth is more likely to be successful.

So first of all, make an assessment of the ground conditions; the soil bearing, porosity, voids, root spread, etc. Secondly, BS 5837 suggests that an arboriculturalist should identify the trees on site, making note of all the species and locations. In reality, many Building Control Bodies do not need an arboricultural report on simple sites – common trees in unproblematic conditions – and so it is worth discussing this early on in the process. Also, be aware that with 150 million urban trees in Britain, a vegetation survey must include reference to conditions

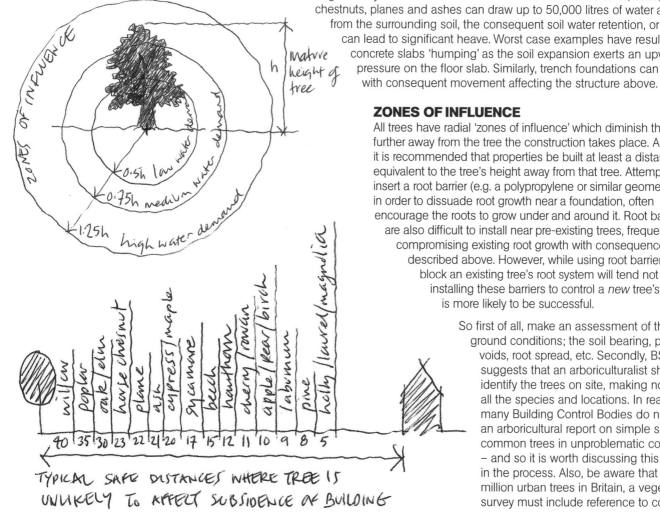

TYPICAL SAFE DISTANCES WHERE TREE IS UNLIKELY TO AFFECT SUBSIDENCE OF BUILDING

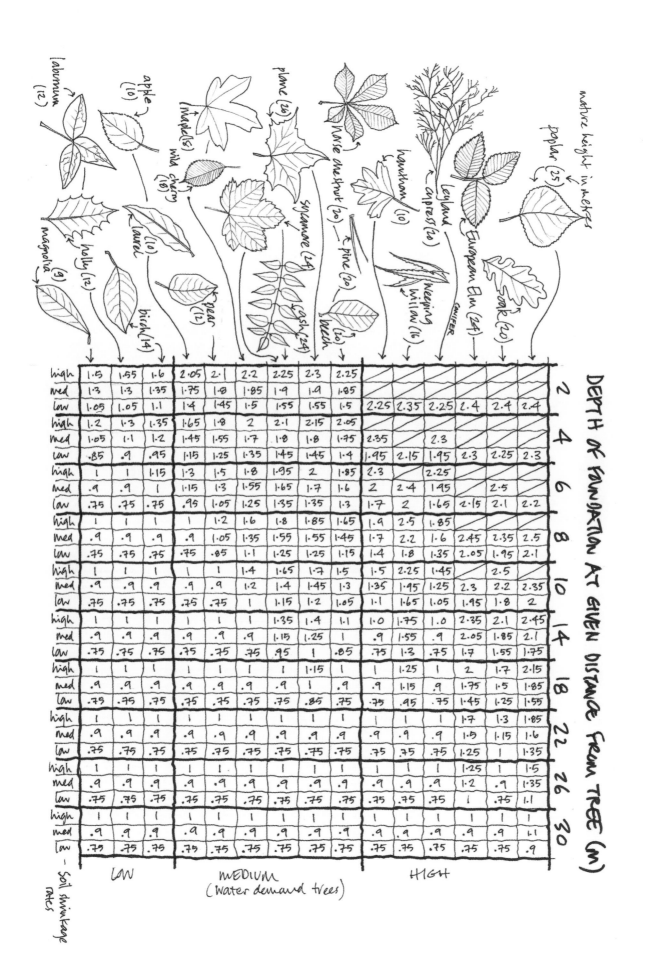

DEPTH OF FOUNDATION AT GIVEN DISTANCE FROM TREE (m)

LOW / MEDIUM (water demand trees) / HIGH

Soil shrinkage rates

> *The removal of trees altogether can leave significant rehydration problems in the soil. Where trees have been recently removed, foundation depths should still be designed to suit the height of the tree.*

beyond the site boundary. Just because you can't see it from your site doesn't mean that it isn't going to affect what you build there. Of the seven million street trees in London alone, it is reputed that 1 per cent have been found liable for damage to properties.

The depth of the foundation that will be necessary to cope with soil shrinkage will depend on the type and proximity of the tree cover. Bear in mind that the trees on site might not have fully matured and it is advisable to take their fully matured height into consideration when planning foundations.

Implementing a policy of pruning or lopping trees to prevent them from growing to their full height, is not a workable long-term strategy for the control of the root network. A crown reduction of 70 per cent by volume (approximately 35 per cent height reduction) reduces the water drawn by only a small amount, and only during the year of pruning. In subsequent years the soil moisture levels return to normal. This is partly because the pruning action encourages shoot growth and hence larger leaves, which then make similar demands on the water absorption capacity of the root system.

The removal of trees altogether can leave significant rehydration problems in the soil. Where trees have been recently removed, foundation depths should still be designed to suit the height of the tree (using the adjoining design table). However, if the removed tree had not reached half the mature height, then the actual height (provided that its height can be stated with some certainty) can be used with NHBC guidelines to assess the foundation depth.

The table provides some indicative safe distances at which a standard 900 mm or 750 mm deep trench foundation is adequate. Where trees give rise to the need for foundation depths of up to and over 2.4 m (shown as a cross in the table), then engineered foundation designs are recommended. The distance between trees and foundation should be inputted into the table; together with information about the relevant height of the tree, and the high, medium or low shrinkage nature of the soils to enable you to read off the foundation depth. Where two or more trees influence the foundation, you should design and specify to the worst case requirements.

The adjoining table is simply a rule-of-thumb guide to foundation depths with a built-in contingency for unforeseen circumstances. Research into the actual, rather than perceived, tree action, the soil 'reaction' and the existing site conditions, can be undertaken to provide real evidence of the site conditions rather than the simplified relationship between vegetation proximity and foundation depth that is drafted here.

Indeed, the specific conditions of actual trees and ground conditions might result in an engineered solution less onerous than the ones quoted in the table. Also, exact surveys ought to reveal problematic conditions such as porous soils overlaying clay soil, underground drainage, water table height and other compensating or exacerbating factors.

References

BS 5837 (2005) 'Trees in relation to construction. Recommendations', BSI.

Crow, P. (2005) 'The Influence of Soils and Species on Tree Root Depth', Forestry Commission.

Department for Communities and Local Government (2006) 'Tree Roots in the Built Environment: Research for Amenity Trees No. 8', DCLG.

Harding, P. (1998) 'How to Identify Trees', Collins.

Hipps N.A., Atkinson, C.J. and Griffiths, H. (2006) 'Pruning trees to reduce water use. Summaries of research – conclusions and recommendations', IP7/06, BRE.

ISE (2001) 'Subsidence of low-rise buildings. A guide for professionals and property owners', 2nd edn, Institution of Structural Engineers.

Reader's Digest (2001) 'Reader's Digest Field Guide to the Trees & Shrubs of Britain'.

RECOMMENDED READINGS

Clark, J. & Matheny, N. (1998) 'Trees and Development – A technical guide to preservation of trees during land development', International Society of Arboriculture.

Lawson, M. (2000) 'Tree Related Subsidence of Low Rise Buildings and the Management Options', Member of the Institute of Biology.

O'Callaghan & Kelly (2005) 'Tree-related subsidence: Pruning is not the answer', Journal of Building Appraisal, Vol. 1, No. 2, pp. 113–129.

National House-Building Council (2006) 'Building Near Trees', Chapter 4.2, NHBC.

Roberts, J., Jackson, N. & Smith, M. (2006), 'Tree roots in the built environment. Research for amenity trees No. 8', DCLG.

04: Flight Simulations
How to pitch a staircase

The National Institute for Health and Clinical Excellence, the organisation with the ironic acronym (Nice) advises architects to 'ensure staircases are clearly signposted and attractive to use... well-lit and well-decorated'.[1] It's got nothing to do with good design; it's yet another insidious campaign to 'reduce obesity'. This Shortcut should be read as part of a calorie-controlled diet.

Building Regulations Approved Documents are regularly reassessed to appraise whether the technical information and scientific efficacy of their guidance remains current and contemporary. A recent report by the Department for Communities and Local Government (DCLG) looks beyond the science. 'The impact of societal changes on the Building Regulations' is a snapshot of modern anthropometric data and is primarily focused on the domestic and residential sector. It signals some proposed changes in Approved Document K: 'Protection from falling, collision and impact' and Approved Document M: 'Access to and use of buildings', although these are not imminent.

The report, written in 2007, has several issues of concern. For the DCLG:

- Dealing with noise transmission at the junction of stairs to party walls is a high priority and it suggests that 'a revision of AD E is appropriate'.

- Stair widths need to be increased and/or a minimum width set.

- The usability of handrails should be improved for an aging population.

- Headroom above internal and external stairs and ramps could possibly be equalised across all Approved Documents at the 2100 mm dimension set in AD M.

- The height of guardings for stairs, landings, ramps and edges of internal floors, could possibly increase to more than the current 900 mm minimum.

Circular handrails should be between 32 mm and 50 mm diameter (between 40 mm and 50 mm for elderly or disabled). Oval handrails should be 18–37 mm x 32–50 mm.

"The term 'dog-leg stair' is frowned upon by British Standards, preferring 'turning stair of two flights with a half-landing'.

The report supports more ergonomic study on a wider range of 21st century humans, to assess the correlations between leg lengths and stride length, so that the requirement for a 400 mm clear landing at the bottom of a flight of stairs, if a door swings across that landing, 'may need investigating to see if it is suitable'. Until such study and subsequent revisions to the Approved Documents appear, the drawings in this Shortcut still stand as current practice.

While AD K is the bible for staircase design in England and Wales, it has not been altered since 2000, whereas many of the aforementioned items have already been taken up in changes introduced between 1 May 2006 and 30 April 2007 in the Scottish Technical Handbooks – see Section 4 'Safety' (STH 4). Additional guidance can be found in Scottish Homes' 'Housing for Varying Needs'.

In STH 4:

- Clause 4.3.1 states that cupboard or duct door swings on landings are permissible provided that they do not reduce the effective width of the landing. (In AD K1, door swings may open onto landings provided that a minimum of 400 mm clear space remains. See also clause 4.3.6.

- Clause 4.2.8 gives guidance on the space to be provided adjacent to a stair flight to accommodate a future stair lift installation, *à la* Lifetime Homes Initiative (see Shortcuts: Book 2).

- Clause 4.3.2 recommends a maximum pitch of any other than a private stair to be 34 degrees and it has simplified the tread/riser calculations significantly. (The nearest reference in AD K is that the maximum pitch for gangways for seated spectators should be 35 degrees.)

- Clause 4.3.15 recommends that a handrail should be fixed at a height of at least 840 mm and no more than 1 m above the pitch line of the stair or landing. (AD K recommends around 900 mm and 1000 mm above the pitch line of stairs, and 1100 mm at landings.) BS 5395-1 suggests that guarding can be reduced to 900 mm in dwellings although its stating that 'this can increase the risk of falls' may be enough to prevent architects taking the risk.

- Finally, clause 4.3.1 states that open risers must have treads that overlap the one below by a minimum of 15 mm. (In AD K1, and BS 5395-1, a minimum of 16 mm is recommended!)

HANDRAILS

Circular handrails should be between 32 mm and 50 mm diameter (between 40 mm and 50 mm for elderly or disabled). Oval handrails should be 18–37 mm × 32–50 mm. Handrails should be 50–100 mm from the face of any guarding or wall. For public or assembly building stairs of over 1.8 m width, the requirement that handrails be provided so that the maximum distance between them is no greater than 1.8 m and no less than 1 m, actually precludes stair widths of 1.8–2 m.

HEADROOM AND CLEARANCE

The headroom (the vertical distance between the pitch and the soffit) over a flight of stairs and landings must be no less than 2 m, and the clearance (that is, the perpendicular distance from the pitch to the soffit) must be no less than 1.5 m. This will be achieved with a headroom of 2 m and a pitch of 41.5 degrees. Note that, in small stair runs of three or four steps, there is a tendency for young children to run and jump to clear them all. In this instance, the headroom should be raised to 1.8 m to reduce the risk of them cracking their heads.

GUARDING/BALUSTRADE

Balusters need to be spaced sufficiently so that a 100 mm sphere can not pass through between them; this is only necessary for stairs 'regularly' used by children under five. Similarly, where under-

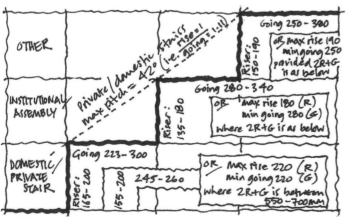

RISERS + GOINGS TO BS 5395-1: 2000 SECTION 7

Approved Document K gives minimum landing / guarding heights as:
- SINGLE FAMILY DWELLING : 900 - 1100 mm (although BS5395-1 states that 900mm "can increase the risk of falls")
- RESIDENTIAL, OFFICES, INSTITUTIONAL, etc: 1100MM
- RETAIL : 1000 mm

In shop or assembly buildings, provide a max 16 risers per flight. Where consecutive flights have more than 36 risers, provide a 30° change of direction along the flight or landing

Where door openings encroach on stair landings ensure a minimum 400mm clear landing space is maintained

min 400

max 42° for domestic stairs

NOTE: All 'treads', 'risers' + 'goings' dimensions may be negotiable for conservation projects and for DDA regulations

min 400mm clear floorspace at foot of stairs

Ensure stair treads provide suitable slip resistance (SEE SHORTCUT 6)

BS5395-1:2000 states that straight stairs may include winders/ 'tapering treads' provided that they only create a ¼ or ½ turn in any given flight

Building Regulations (England + Wales) have NO minimum width for domestic stairs although it is advisable to check with AD B : FIRE and local Lifetime Homes proposals. In Scotland , minimum 800MM (plus handrail ?) is recommended

min 2m clearance above pitch (unless loft access)

2m

max / min goings (see table) measured at 270 mm from outer edges

> 1m

270   270

less than 1m

In flights with winders, measure goings along the centre line of treads

The width of stair flights tends to be the distance between the innermost faces of handrails

There should be no fewer than 3 risers in any one flight , with minor exceptions

NB: Helical stairs describe a helix around a void
Spiral stairs describe a helix around a central column

In helical / spiral stairs , the requirement for 2m headroom does not apply. Also the number of risers can rise to 22 in a given flight provided that a landing is not practicable

Handrails to be 900 - 1000mm above pitch line and if staircase is more than 1m wide , a handrail is required on both sides (for straight as well as curved stairs)

NB: If stair flight width is greater than 1.8m , a middle handrail is required

Where children under 5 years old are regularly present, guarding and balustrades must pass the 100mm diameter sphere test

min 50mm

The length of landing at head of stairs to be at least equal to the width of flight

Balusters (NOT "Banisters")

The goings (+widths) of winders (NOT called "tapered treads") measured along the centre line must be no less than the going (and width) of the treads of the straight flight

NB: The words 'banister' and 'tapered treads' are 'deprecated'

Any perforations in steel treads must not exceed 20×20mm (or 20mm clear gap in slatted treads)

GLOSSARY OF TERMS

Private stair – Main access stair located in a private dwelling or similar situation, and which is used by a small number of people who are generally familiar with it.

Public stair – Stair used intermittently by a large number of people, some of whom are not familiar with it, and which is located in a commercial building, office or public building such as a library.

Assembly stair – Stair used simultaneously by a large number of people, many of whom are not familiar with the stair, and which is located in an assembly building such as a theatre, concert hall, educational establishment or stadium.

Bulkhead – The soffit of a ceiling above a stair and usually constructed parallel to its pitch line.

Overlap – Amount by which the nosing of a tread (including a landing) oversails the next lower tread (or landing).

Patent nosing – Material added to the leading edge of a tread, usually to increase its slip-resistant properties.

Raked riser – Riser which slopes from the rear of a tread towards the nosing of the next tread (or landing) above it.

Simply supported stair – Stair in which both ends of each step are supported.

Cantilevered stair – Stair in which one end only of each step is supported. (A double cantilevered stair is one supported only on a carriage, usually centrally positioned.)

Turning stair – Stair in which the direction is changed. (A double return stair is one that has one flight to a landing and two change-of-direction flights springing from that one landing.)

Dog-leg stair – This term is frowned upon by British Standards, preferring 'turning stair of two flights with a half-landing'.

Geometrical stair – Turning stair in which the outside string and handrail continue in an unbroken line with curved ends.

Winding stair – Turning stair that includes winders.

Spiral stair – Turning stair that describes a helix around a central column.

Helical stair – Turning stair that describes a helix around a stairwell.

fives are regularly present (or to prevent the elderly feeling 'a sense of insecurity when looking through the spaces between treads') open risers shall not have an accessible gap that will permit the passage of a 100 mm diameter sphere. This means that, in the case of, say, a 220 mm riser, the underside of the tread ought to be provided with a downturn angle, or a similar solution, to reduce the open riser dimensions.

The Scottish Technical Handbook states that in domestic premises a handrail need only be fitted on one side of a staircase, provided that 'the side on which the handrail is not fixed should permit the installation of a second handrail at a future date'. Given that the minimum domestic stair clear width dimension is 800 mm and a secondary handrail ought to project 50–100 mm from the guarding line, de facto staircase widths are now 850–900 mm. (There is no minimum domestic stair width given in the England and Wales Approved Documents.)

WINDERS

Even though they are called 'tapered treads' in the Approved Documents, British Standards frown upon the term, preferring them to be called 'winders'. Winders in stair widths up to 1 m should be measured along the centre line of the flight as described in BS 585-1. For winders of greater width, the going of each winder should be measured at a distance of 270 mm from the inner and outermost edges. While AD K states that in helical and spiral stairs the minimum going should not be less than 50 mm, BS 5395-1 states that it should, where possible, be no less than 75 mm. Assembly building stairs should not contain winders.

[1] Watts R. and Hennessy P., 'Make fat people use the stairs, architects told', Daily Telegraph, 5 November 2007.

References

Department for Communities and Local Government (2007) 'The Impact of Societal Change on the Building Regulations', Final Report: BD 2512, DCLG.

Pickles, J., Scottish Homes (1998) 'Housing for varying needs; A design guide: Part 1 Houses and flats', TSO.

RECOMMENDED READINGS
BS 5395:1 (2000) 'Stairs, ladders and walkways. Code of practice for the design, construction and maintenance of straight stairs and winders', BSI.

BS 5395:2 (1984) 'Stairs, ladders and walkways. Code of practice for the design of helical and spiral stairs' (incorporating amendment No. 1 and Corrigendum No. 1), BSI.

Office of the Deputy Prime Minister (2000) 'The Building Regulations 2000: Approved Document K: Protection from falling, collision and impact', NBS.

Office of the Deputy Prime Minister (2004) 'The Building Regulations 2000: Approved Document M: Access to and use of buildings', NBS.

Scottish Executive (2007) 'Technical Guidance. Section 4: Domestic: Safety', SE.

05: A Dangerous Oversite
Sulfate attack

Warning! This Shortcut contains severely degraded, hardcore material. May cause unsightly swelling. Exposed cracks could cause offence. Reference to colliery slags has been used under advisement. Early evidence of adverse reactions should be reported; heaving requires specialist attention. Protective membranes should be used at all times.

Hardcore is described in BRE Digest 276 as 'a make-up material to provide a level base on which to cast a ground-floor slab, to raise levels, and to provide a dry, firm base on which work can proceed or to carry construction traffic'. Fill can be divided into non-loadbearing (placed below soft landscaped areas), and loadbearing (below buildings, roads and most hard landscaping). Non-loadbearing fill may still need engineering input to ensure stability where slopes are involved.

Hardcore fill under concrete slabs has been commonly used in domestic ground-floor construction since the 1940s; before that, suspended timber joists were the usual way of raising ground floor levels. In 1951, the Ministry of Housing Manual stated that 'due to the present shortage of timber, the construction of ground floors of wood joists and boards is not allowed except on steeply sloping sites where the cost of a solid concrete floor would be prohibitive. Solid ground floors, which are being laid generally today are proving highly satisfactory when properly constructed and seem likely to be used even if timber becomes freely available again.'[1]

Hardcore fill under concrete slabs has been commonly used in domestic ground floor construction since the 1940s; before that, suspended timber joists were the usual way of raising ground floor levels.

> *Ordinary Portland Cement (OPC) is susceptible to sulfate attack; therefore, if it is known that subsoil or hardcore values are above 5.5 pH, OPC must not be used.*

The two most significant hazards relating to hardcore fill material under concrete floor slabs are settlement (frequently caused by poor compaction or the wrong type of material) and chemical damage caused by a reaction between the fill and the overlying concrete. This Shortcut concentrates on the latter.

SETTLEMENT

Back in the 1970s it was common for hardcore fill under houses to be around 1.5 m deep and more. Nowadays, the National House Building Council's 'NHBC Chapter 5.1' advises that ground bearing floors may not be suitable, where (amongst other things) the 'thickness of the fill is more than 600 mm'. A civil or structural engineer should approve fill material of a greater depth.

Hardcore needs to be properly graded (i.e. with a good mixture of large and small pieces which interlock to give strength), be laid on a firm formation level and be thoroughly compacted in layers not exceeding 150 mm deep.

Surfaces of excavations with a gradient greater than 1:5 (20 per cent) to receive fill will generally require horizontal benches – stepped excavations – cut to match the depths of compacted layers of filling. Note that the Health and Safety Executive have outlawed the phrase 'The Angle of Repose', replacing it with 'The Maximum Allowable Slope'.

CHEMICAL ATTACK

Concrete floor slabs and oversite concrete (i.e. the concrete laid beneath suspended floors) has been particularly susceptible over the years to attack by water-borne sulfates in hardcore. The problem arises from the migration of sulfates, in damp conditions, into the underside of the oversite/concrete and their reaction with the constituents of the concrete. Known as 'sulfate attack', it causes the concrete to break down, expand, lift and crack.

The presence of sulfates can be due to 'naturally' occurring sulfate-bearing hardcore material such as burnt colliery shale (red ash or red shale), furnace bottom ash (black ash), blast-furnace slag, oxidised pyretic shales, or from demolition rubble being contaminated by substantial quantities of gypsum, for example. Approved Document C (AD C) warns of geographic areas where sulfates are likely to be found. It doesn't show where these locations are – see map below – but concrete in contact with soils in these locations, or hardcore derived from them, should be checked. The common sulfate compounds to look out for are:

- calcium sulfate
- magnesium sulfate
- sodium sulfate
- potassium sulfate.

Calcium sulfate – frequently occurring as gypsum and anhydrite – is the most prevalent in UK soils but is the least harmful to concrete. Recycled hardcores may accidentally include gypsum plaster. Other hardcore shale may have sulfate content which appears to be safely 'fused' into the material but which can be released and become soluble in situ.

The main factors exacerbating the risk from sulfates are:

- the amount of sulfate present
- the sulfate's ability to migrate into the concrete, i.e. moisture movement.

Ordinary Portland Cement (OPC) is susceptible to sulfate attack; therefore, if it is known that subsoil or hardcore values are above 5.5 pH, OPC must not be used. Sulfate attack occurs when moisture from the subsoil/hardcore enters the concrete, causing sulfide salts (and chlorides) to crystallise as the water evaporates. As more and more crystals are deposited, the forces acting on the concrete cause it to fracture or to bow. A period of several years is normally needed before damage becomes visible.

This is the most common form of sulfate attack on concretes containing OPC and is known as 'ettringite'. It develops high stresses and cracking in the concrete. White crystalline deposits are the tell-tale signs of this conventional form of sulfate attack.

LIKELY INCIDENCE AREAS FOR SULFATE-BEARING HARDCORE

A second form of damage, known as the 'thaumasite form of sulfate attack' (TSA), was only discovered in the 1990s in the UK. It requires particular conditions, such as:

- a great deal of water-soluble sulfates

- a limestone aggregate

- a non-carbonated concrete (or bicarbonate in the ground water)

- persistent wetness

- temperatures below 15°C.

TSA is a reaction within the concrete mix itself. Limestone is a source of calcium carbonate which reacts with moisture-borne sulfates such that the cement matrix – the strength-giving properties of the concrete – are eroded, causing the concrete surface to soften and eventually to disintegrate. Oolitic limestone has been shown to be the most reactive, and investigations by the BRE have described concrete affected by TSA as having turned to 'mush', even when sulfate-resisting Portland cement was used. In 2005, the British Cement Association stated that it had identified only 60 instances of TSA. That same year, BRE's Special Digest 1 was developed to include consolidated knowledge on the subject of TSA.

While the impact of sulfate attack can be devastating – with the concrete effectively being 'eaten away' or severely bowed – the dangers of sulfate attack only arise where an OPC concrete slab is in direct contact with the subsoil/hardcore. Damage can be halted simply by the inclusion of a dpm to separate the two materials. In effect, since domestic buildings constructed since the 1970s have tended to incorporate dpms as a matter of course – problem solved. In fact, the Department for Communities and Local Government (DCLG) states that domestic buildings constructed from the early 1970s onwards are 'unlikely' to have concrete floor slabs affected by sulfate problems. Earlier buildings may not be so lucky.

Trench foundations and pad footings, located as they often are, near the natural drainage points for hardcore, are also liable to attack (but are not mentioned in the DCLG's latest remediation advice report, 'Sulfate damage to concrete floors on sulfate-bearing hardcore'). Strip footings, for example, may shear along the line of a supported loadbearing wall if sulfates are allowed to weaken the concrete. In these circumstances, sulfate-resisting cement in the concrete is usually specified. Sulfate-resisting Portland cement (SRPC) is suitable for foundations and certain concrete applications where groundwater has a concentration of SO_3 up to 0.1 per cent, or subsoils with up to 0.5 per cent SO_3. The useful book, 'Building in the 21st Century' states that this cement is commonly specified for foundations in clay or gravel areas, such as Essex and London, where pockets of sulfates are present[1].

Aside from isolating the concrete from the subsoil/hardcore – say, using a 300 μm dpm – sulfates will not present a problem if the subsoil/hardcore remains dry.

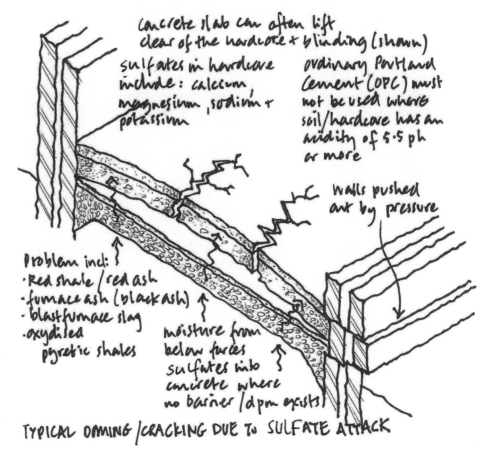

concrete slab can often lift clear of the hardcore + blinding (shown)

sulfates in hardcore include: calcium, magnesium, sodium + potassium

ordinary Portland cement (OPC) must not be used where soil/hardcore has an acidity of 5.5 ph or more

walls pushed out by pressure

Problem incl:
· Red shale / red ash
· furnace ash (black ash)
· blastfurnace slag
· oxydised pyretic shales

moisture from below forces sulfates into concrete where no barrier / dpm exists

TYPICAL DOMING / CRACKING DUE TO SULFATE ATTACK

The power of the specification

In the case of Rotherham Metropolitan Borough Council v Frank Haslam Milan (1996) [59 Con LR 33[2]], an architect specified granular hardcore (in some detail), including the description of the hardcore to be of 'graded or uncrushed gravel, stone, rock fill, crushed concrete or slag'.

The non-specialist contractors used steel slag – a material which is subject to expansion in confined conditions – resulting in cracking of the concrete floor. They claimed ignorance of the detrimental impact that steel slag would have, even though they had a duty of care to lay a floor that was fit for purpose, which they did not do.

However, it was held that the specification clause categorically confirmed that slag would be acceptable, albeit not steel slag, but the architect had the opportunity to have made the distinction, which it did not do. The contractor was found not liable.

Cited in Furmston, M. (2006), 'Powell-Smith & Furmston's Building Contract Casebook', Blackwell Publishing.

RECYCLE IN HASTE

The UK government's Waste and Resources Action Programme (WRAP) ties into an 'AggRegain' campaign to promote the Aggregates Programme (funded by the Aggregates Levy Sustainability Fund – see Shortcuts: Book 2). This aims to reduce demand in England for primary materials by encouraging the use of aggregates from recycled and secondary resources. The Landfill Tax also imposes a charge on excavated material – the material replaced by the imported fill. However, on many small and medium-sized projects, imported hardcore is still cheaper than the processing and testing that would be needed for inferior quality reuseable material excavated from site.

If you plan on reusing excavated material for loadbearing purposes, a site investigation report should reveal the likely quantity and quality available, but it will still require diligent site supervision to check that each batch is up to the required standard. Reused material may require processing on or off site (e.g. a site crusher may be employed on large projects) and at the very least it will require careful compaction in layers to ensure the correct loadbearing capacity. The economics of reprocessing and reusing excavated material versus the use of imported fill materials, which themselves are of varying quality and cost, will to some extent only become apparent as the excavations proceed and are likely to involve proposals and decisions by the contractor, consulting engineer, or both.

However, as an example of the need for caution in the headlong rush towards contemporary recycling, it is worth noting that the government in the 1930s promoted the use of blast-furnace slag and burnt colliery shale as approved materials for use as hardcore. These waste materials, a common sight in the slag heaps and spoil tips of Britain's interwar industrial landscapes, were recklessly appropriated for use in the construction industry in order to find a suitable recycling outlet for those waste products. Over the course of the intervening years, we have come to realise that this was a mistake, as these slags – combined with historic naïve detailing – resulted in chemical corrosion of the concrete.

[1] Ministry of Works: Ministry of Local Government and Planning. 'Housing Manual 1949. Technical Appendices', HMSO quoted in Longworth, I. (2008), 'Sulfate damage to concrete floors on sulfate-bearing hardcore', DCLG, p16.

[2] Legal references explained in Shortcuts: Book 2

References

BS 812-111 (1990) 'Testing Aggregates. Methods for Determination of Ten Per Cent Fines Value (TFV)', BSI [partially replacing BS EN 1097-2:1998].

BS EN 1097-2 (1998) 'Tests for Mechanical and Physical Properties of Aggregates. Methods for the Determination of Resistance to Fragmentation', BSI [partially replacing BS 812-111: 1990].

Building Research Establishment (2001) 'Concrete in Aggressive Ground: Assessing the Aggressive Chemical Environment' (amended March 2003), (no longer current but cited in Building Regulations), Special Digest 1 Part 1, BRE.

Building Research Establishment (2001) 'Concrete in Aggressive Ground: Specifying Concrete and Additional Protective Measures' (amended March 2003), (no longer current but cited in Building Regulations), Special Digest 1 Part 2, BRE.

Building Research Establishment (2001) 'Concrete in Aggressive Ground: Design Guides for Common Applications' (amended March 2003), (no longer current but cited in Building Regulations), Special Digest 1 Part 3, BRE.

Building Research Establishment (2001) 'Concrete in Aggressive Ground: Design Guides for Specific Precast Products' (amended March 2003), (no longer current but cited in Building Regulations), Special Digest 1 Part 4, BRE.

Card, G.B. (1995) 'Protecting Development from Methane: Methane and Associated Hazards to Construction', CIRIA Report 149, Construction Industry Research and Information Association.

Cooke, R. (2007) 'Building in the 21st Century', Blackwells.

Humphries, Dr R.N. (2000) 'Good Practice Guide For Handling Soils, Sheet 18: Soil Decompaction by Excavator Bucket', Farming and Rural Conservation Agency, Cambridge.

RECOMMENDED READINGS

Building Research Establishment (2008) 'Sulfate damage to concrete floors on sulfate-bearing hardcore: identification and remediation', DCLG.

Building Research Establishment (2005) 'Concrete in Aggressive Ground', 3rd edn, Special Digest 1, BRE.

Building Research Establishment (2003) 'Floors and Flooring, Performance, Diagnosis, Maintenance, Repair and the Avoidance of Defects', Building Report 460, BRE.

Longworth, I. (2008) 'Sulfate Damage to Concrete Floors on Sulfate-bearing Hardcore', DCLG.

National House Building Council (2007) 'NHBC Standards Part 5 – Substructure and Ground Floors', NHBC.

Office of the Deputy Prime Minister (2004) 'Approved Document C: Site Preparation and Resistance to Contaminants and Moisture', ODPM.

06: Damp-proof Courses
A short introduction to rising damp

The National Statistics Office claims that one in every ten over-50-year-olds complains of damp rising in floors and walls. But records of peoples' perceptions can be misleading given that damp appears to have been all but eliminated in modern properties. This Shortcut looks at good damp-proof detailing in traditional construction.

The discovery of rising damp in a property inevitably results in hefty quotes by specialist contractors, panic by the homeowners, followed by months of inconvenience as the remedial work is carried out. However, rising damp is one of those catch-all phrases that needs clarification.

The victims of the recent floods in the UK, for example, undoubtedly returned home to find the flood water soaking into the walls and rising even higher than the nominal water level. However, the Building Research Establishment (BRE) says that this sort of damage usually cannot be counted in the rising damp statistics. Rising damp occurs when relatively porous walls are built in saturated soil, causing moisture to rise by capillary action. Where the soil is not saturated, the capillary action is countered by the suction of the soil. In cases where the soil suction is greater than the capillary pressure, no rising damp will occur. A correctly specified and installed damp-proof course (dpc) can intervene to prevent capillary action and stop the rise of moisture up the wall. In truth, the best dpc in the world will not avert flood damage.

Always look out for leaking pipes, blocked gullies, spilled dog bowls – before diagnosing rising dampness.

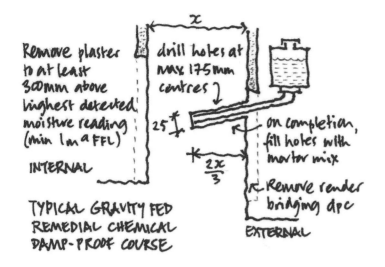

Remove plaster to at least 300mm above highest detected moisture reading (min 1m a FFL)

INTERNAL

drill holes at max 175mm centres

25

2x/3

on completion, fill holes with mortar mix

Remove render bridging dpc

EXTERNAL

TYPICAL GRAVITY FED REMEDIAL CHEMICAL DAMP-PROOF COURSE

" It is easy to misconstrue damp meter readings. A misleading diagnosis may arise from the presence of hygroscopic salts, for example.

SOME MECHANISMS OF MOISTURE INGRESS :-

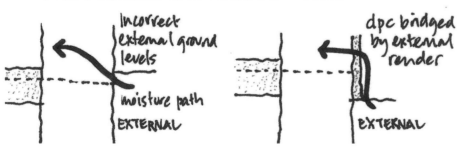

Incorrect external ground levels

moisture path

EXTERNAL

dpc bridged by external render

EXTERNAL

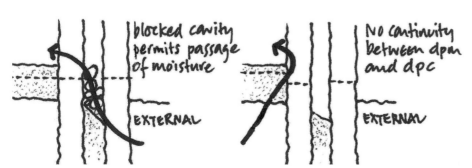

blocked cavity permits passage of moisture

EXTERNAL

No continuity between dpm and dpc

EXTERNAL

In many instances, dpcs are also installed to prevent downward moisture, rainwater, etc. from entering the building and spreading onto the internal surfaces. This Shortcut only explores the means of protection against rising damp.

The most obvious tell-tale sign of rising damp is consistent darkening of plaster or wall coverings, indicating the prevalence of excessive moisture. This is accompanied by a tide-mark at approximately 1–1.2 m height which is often demarcated by a line of salt crystals deposited as the moisture evaporates – these tend to be deposits of nitrates and chlorides from the soil. Dampness arising from external hard surfaces (a blocked gulley causing ponding in an adjoining concrete yard, for example) does not generally contain soil salts but might also give rise to a crystalline appearance on inner or outer surfaces due to the leaching out of the natural sulfates contained within the bricks themselves. In many circumstances, sending samples for analysis is the only way to determine whether the problem is rising damp.

It is important then, that those inspecting for evidence of damp should also use a commonsense approach. Always make diagnostic observations – look for leaking pipes, blocked external gullies, spilled dog bowls, etc – before declaring the area suffering from rising dampness. Also, when testing for damp in a property, remember that the ubiquitous damp meter actually measures electrical resistance and so high readings taken from a wall do not necessarily indicate the presence of dampness. It could be that there is insufficient plaster coverage over a metal pipe, for example; maybe the foil backing to plasterboard has been badly laid and exposed. A misleading diagnosis may also arise from the presence of hygroscopic salts within the wall. Hygroscopic salts absorb moisture from the atmosphere and the greater the prevalence of these salts, regardless of the amount of actual moisture in the wall, the greater the electrical conductance. It is easy to misconstrue meter readings.

BRE Report 466 'Understanding dampness' contains a very useful tip for working out whether evidence of moisture on a given surface is caused by condensation or rising damp. Taking a floor slab as a case study, it advises that you 'place a piece of (kitchen) foil about half a metre square on the floor under any carpet/underlay and seal it firmly round the edges with adhesive tape. Inspect it the next day: if moisture has collected on the underside of the foil, there is dampness in the slab. If the moisture is on the upper surface, it is condensation.'

Rising damp occurs predominantly in older buildings (especially those with solid external walls) and those with solid floors installed without dpms (which was common practice in the 1950s and 1960s). Staged remedial work to these properties in the 1970s and 1980s frequently resulted in the wall's dpc being bridged by new concrete floors. Ironically, this often encouraged the passage of more ground moisture into the walls than before. Where plaster is in contact with the floor, the moisture is more likely to rise than if its passage relied solely on moisture movement up the brickwork or blockwork. Other common unintended consequences of remedial damp-proofing include the accidental blockage of underfloor vents preventing through-ventilation from adjoining suspended floor areas.

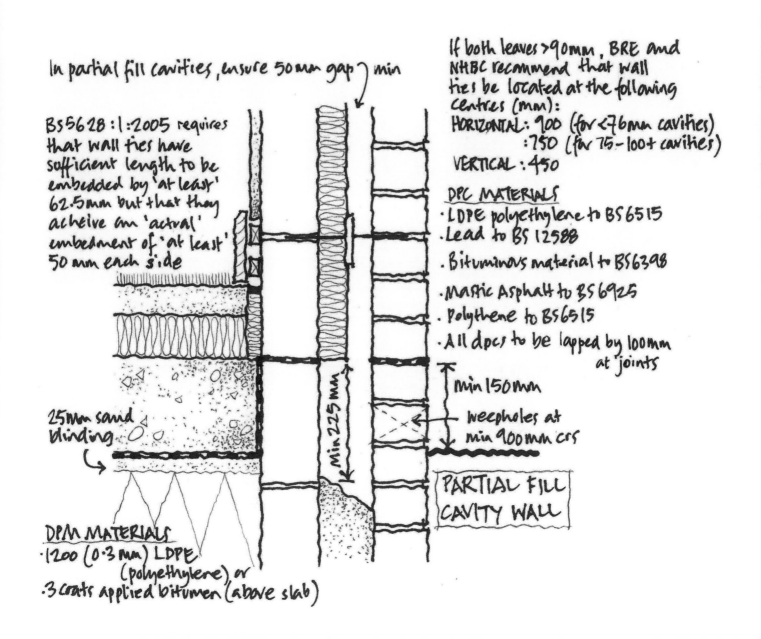

In partial fill cavities, ensure 50mm gap ⌐ min

BS 5628 :1 :2005 requires that wall ties have sufficient length to be embedded by 'at least' 62.5mm but that they acheive an 'actual' embedment of 'at least' 50mm each side

25mm sand blinding

Min 225 mm

DPM MATERIALS
· 1200 (0·3mm) LDPE (polyethylene) or
· 3 coats applied bitumen (above slab)

If both leaves >90mm, BRE and NHBC recommend that wall ties be located at the following centres (mm):
HORIZONTAL: 900 (for <76mm cavities)
: 750 (for 75-100+ cavities)
VERTICAL : 450

DPC MATERIALS
· LDPE polyethylene to BS 6515
· Lead to BS 12588
· Bituminous material to BS 6398
· Mastic Asphalt to BS 6925
· Polythene to BS 6515
· All dpcs to be lapped by 100mm at joints

min 150mm

weepholes at min 900mm crs

PARTIAL FILL CAVITY WALL

In over 45,000 dwellinghouses in England, rising damp affected just 3.9 per cent of dwellings.

EFFECTIVE BARRIERS

Simply put, preventing moisture migration is the role of the dpc, an impermeable barrier to the passage of moisture. Scottish Technical Handbook 3, clause 3.4 refers to the dated Code of Practice CP102 (1973). Its central requirements match those of Building Regulations (E&W) Approved Document C. This describes four requirements for effective damp-proof courses which, notwithstanding minor variations, can be synopsised as follows:

1. The dpc should be impervious and continuous.

2. In external walls, dpcs should be at least 150 mm above adjoining ground.

3. In external cavity walls, the cavity should extend 225 mm below the level of the dpc, and weepholes be provided at 900 mm horizontal centres.

4. Cavity trays that do not extend the full length of a wall (e.g. over a meter box opening at low level, for example) should have stop ends and weep holes to prevent run-off within the cavity from rainwater ingress above. BS 8215: 1991 adds that 'trays that bridge a cavity should be stepped by not less than 150 mm from the outer to the inner leaf'.

There are circumstances in which certain materials, which otherwise comply with the above requirements, are inappropriate.

For example:

- Polyethylene sheet is not suitable if only lightly loaded, i.e. under copings, etc., where slippage of the element above the dpc may occur.

- Lead will corrode in contact with mortar and must be painted on both sides with bituminous paint.

- Slate dpcs are discontinuous and should be formed in two half-lapped courses.

- Existing walls insulated with polystyrene insulation should not be remedially treated with chemically injected solvent-based products.

If you've missed out on any of these, before you reach for your insurer's telephone number to cover yourself for possible actions, it is worth remembering that the risks are low. In 1996, the English House Condition Survey (EHCS) – a survey of around 45,000 dwellinghouses (with the results extrapolated to cover all of England) – confirmed that rising damp affected just 3.9 per cent of dwellings and penetrating damp 6 per cent. By 2006, rising damp had been further restricted to 2 per cent of the housing stock, and penetrating damp to 5 per cent. In the Scottish House Condition Survey 2004/05, 'rising or penetrating damp' was located in just 5 per cent of properties. As the older stock gets replaced by better quality new construction, it's not hard to imagine a time in the very near future when rising damp will effectively be eliminated from your insurer's portfolio.

References

BS 6576 (2005) 'Code of practice for diagnosis of rising damp in walls of buildings and installation of chemical damp-proof membranes', BSI.

BS 6515 (1984) 'Specification for polyethylene damp-proof courses for masonry', BSI.

Harrison, H., Sanders, C. & Trotman, P. (2004) Report 466 'Understanding dampness – effects, causes, diagnosis and remedies', BRE Press.

Office of the Deputy Prime Minister (2004) 'Approved Document C: Site preparation and resistance to contaminants and moisture', NBS.

RECOMMENDED READINGS

Billington, M.J. (2007) 'Using the Building Regulations. Part C: site preparation and resistance to contaminants and moisture', Elsevier Butterworth-Heinemann.

BS 5628-3 (2005) 'Code of practice for use of masonry – Part 3: Materials and components, design and workmanship', BSI.

BS 8215 (1991) 'Code of practice for design and installation of damp-proof courses in masonry construction', BSI.

BS 8102 (1990) 'Code of practice for protection of structures against water from the ground', BSI.

Building Research Establishment (2007) Digest 245: 'Rising damp in walls. Diagnosis and treatment', BRE.

07: Movement Joints
A flexible approach to masonry

Typically, clay bricks tend to expand due to their absorption of water over time, whereas concrete tends to dry out and shrink over similar periods. This expansion and contraction is also affected by exposure, elevation and climate. Suffice to say that allowances must be made for the fact that different materials move differently in buildings.

Clay bricks often arrive on site, kiln-dried and ready to reabsorb moisture; aircrete/aggregate concrete blocks, on the other hand, tend to retain moisture during their steam-pressurised autoclave curing process and are more likely to shrink in normal atmospheric conditions. Kiln-dried structural timber arriving on site with a moisture content of 20 per cent may swell as it absorbs moisture or shrink if fitted in heated, unventilated spaces. The 'excess' water used in a concrete mix to give it suitable workability qualities will gradually evaporate causing concrete shrinkage, resulting in possible cracks in the concrete itself, or gaps between the concrete and surrounding materials.

From the diurnal variations in cladding materials responding to the fluctuations between day-and night-time temperatures to the long-term gradual dimensional changes of some structural materials, all buildings are subject to movement and allowances must be made to accommodate it where possible and practicable. Fired clay wall and floor tiles, for example, especially if used in warm, wet areas, have a tendency to expand significantly, and so a soft sealant joint, as opposed to grout, should be included at 3–4.5 m horizontal and vertical centres. Also, as recommended in BS 5385, these movement joints should pass right through the thickness of tile, bedding and rendering to the background, and the tiles must not bridge structural movement joints.

Normal storey height-masonry walls expand by around 1 mm/m over the lifetime of the building and this is an irreversible process.

Here we examine the role of movement joints, including expansion and contraction joints (but not assembly joints or slip planes) in masonry.

A movement joint is a break in the structure that has been designed to accommodate dimensional changes in the materials that make up that structure or structural element, but is equally appropriate to non-structural applied materials, like render or tiling. Where large areas of clay bricks are used in a wall, for instance, the expectation is that they will expand cumulatively, increasing the overall length of the wall, resulting in pressure on neighbouring structures or on elements that will not move in a similar fashion. BS 5628 'Code of practice for use of masonry', assumes that normal storey-height masonry walls expand by around 1 mm/m over the lifetime of the building and this is an irreversible process, i.e. the bricks do not contract back to their original dimensions at some point. This rate of expansion can be minimised if the walls are restrained, by buttresses, floor joists, straps, etc., but as far as the NHBC is concerned, movement joints in clay brick walls must be at maximum intervals of 15 m (or a maximum 10 m in unrestrained situations such as parapet walls, etc.).

The movement joint effectively breaks the wall into manageable panel sizes, distinct and separate to each other, yet tied in to the remainder of the wall (or structure) for stability and continuity. In this way, the cumulative expansion can be handled more easily and sensibly. The width of a joint (in mm) should be about 30 per cent more (in numerical terms) than the distance between joints (in metres). So if the distance between joints is 15 m, the joint should be (15 + 5) = 20 mm wide (although given that this is the maximum spacing, guidance should be obtained from manufacturers to ensure that the chosen sealant is suitable). Horizontal movement joints in unrestrained walls should be at the same intervals as vertical ones.

In domestic properties, movement joints are not normally required for inner leaf blockwork because the 6 m dimension at which contraction becomes a significant problem will seldom be reached. In larger buildings, however, in order to accommodate shrinkage in blockwork panels in excess of the 6 m dimension, a 10 mm movement joint should be provided and filled with weak mix mortar, or using hemp, fibreboard or cork as filler materials. In contraction joints, these board materials are simply used for their relative durability, i.e. they are only used to define the gap as the building works continue and to provide a backing against which a sealant can be inserted; they do not expand to fill the widening joint gap. Conversely, they have unsuitable compression characteristics to be used in expansion joints where the filler should be a flexible cellular polyethylene or polyurethane, or a foam rubber material that can easily be compressed to 50 per cent of its original thickness. Any filler material fitted in a movement joint should be installed such that it maintains a recess from the face of the wall (both faces if in a parapet or similar) to accommodate the water-resistant sealant such that width:depth ratio is 2:1 or 1:1 (for elastoplastic sealants, including one- and two-part polysulfides) or 1:1 or 1:2 for plastoelastic sealants (see later section).

Diagram labels: >6m; potential 6mm expansion; <675mm; tendency to crack; possible mechanical couple formed causing rotation and cracking at joints; compressible seal; slide by joint

	clay	calcium silicate	conc. brick or block	natural stone	parapet walls
Joint width (mm)	16	10	10	10	10
usual joint spacing (m)	12	7.5 – 9	6	15 – 20	half previous figures
Additional information	15 m max spacing	max length:height = 3:1 compressible seals at max 30 m crs		max 8m from corners	max 1·5m from corners

Where possible, movement joints should be located at changes in direction or of thickness in a wall, and start at least one half of the recommended centre to centre dimension from a return. They should also run the full height of the element and joints in the substructure must be carried through into the superstructure. In traditional cavity construction, wall ties which would otherwise be fixed at 900 mm horizontal and 450 vertical centres will be reduced to 300 mm in both directions surrounding the movement joint.

If there is a return to a movement joint in a clay masonry wall of 675 mm or less and one of the adjoining lengths of brickwork wall is more than 6 m (see diagram opposite) then there is the possibility that the pressure exerted by the long wall may rotate the return brickwork. In this instance, a compressible seal or a 'slide by' joint is required. In a calcium silicate and concrete masonry wall, because both materials are prone to contract, its length:height ratio should not exceed 3:1. BS 5628-3: 2005 states that 'external walls of concrete (and calcium silicate) masonry should have a compressible joint to allow for thermal expansion at not more than 30 m intervals.'

Timber tends not to expand and contract along its length to any significant degree, but will swell and shrink in its cross sectional area. Therefore, differential movement between the leaves of brick-clad/timber-framed structures is most noticeable where significant amounts of cross-grain timber are used, such as in sole and header plates. For this reason, composite sole plates formed from multiple laminates of timber are not recommended, and soft joints between the timber and structure (6 mm on ground floors, 12 mm on first floors, etc.) are recommended. Accordingly, other timber plates – such as those at window/door cills and heads – need 3-mm-thick soft joints on the ground floor, 9 mm on the first floor, etc.

Wall ties play an important stabilising role across movement joints and need to be firmly bedded or fixed in place. BS EN 845-1 'Specification for ancillary components for masonry' covers various types of masonry ties (superseding DD 140-2 (1987) 'Wall ties. Recommendations for design of wall ties'). Cavity wall ties are classified into asymmetrical or symmetrical, horizontal, slope-tolerant or movement-tolerant, the latter being defined as a 'cavity wall tie which is designed to allow large

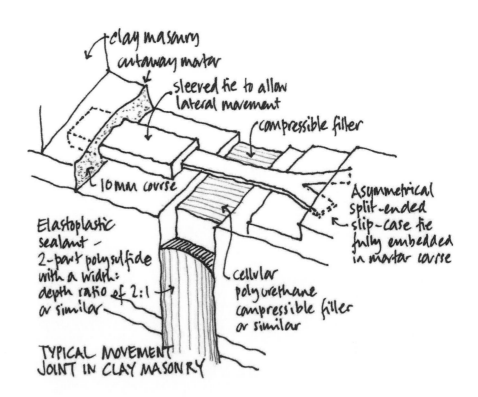

clay masonry
cutaway mortar
sleeved tie to allow lateral movement
compressible filler
10 mm course
Asymmetrical split-ended slip-case tie fully embedded in mortar course
Elastoplastic sealant – 2-part polysulfide with a width:depth ratio of 2:1 or similar
cellular polyurethane compressible filler or similar

TYPICAL MOVEMENT JOINT IN CLAY MASONRY

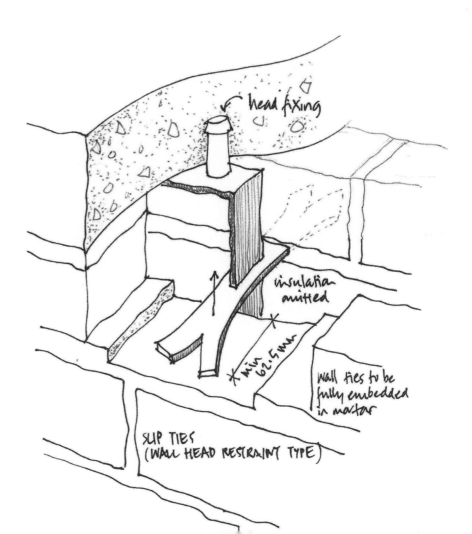

head fixing
insulation omitted
min 62.5 mm
wall ties to be fully embedded in mortar

**SLIP TIES
(WALL HEAD RESTRAINT TYPE)**

"

Concrete (and calcium silicate) masonry should have a compressible joint to allow for thermal expansion at not more than 30 m intervals.

in-plane differential movements of the walls, without generating large shear stresses, by the use of flexible body materials, free-running slot systems, articulated joints or other means.'

In masonry cavity walls, double fishtail ties are not appropriate across movement joints as at least one end must be 'sleeved' in order that the masonry can move around it without dislodging the tie embedment. These slip ties are divided into 'general purpose' and 'wall head restraint'. General purpose slip ties allow the transmission of shear forces between two adjacent sections of masonry, say, while allowing in-plane movement. Head restraint slip ties, also known as sliding anchors, restrain a masonry panel to a horizontal structural element such as a floor or roof slab, but allow differential vertical movement between the masonry and the structural element.

BS EN 845 covers a range of wall tie materials, from copper to polypropylene. Stainless steel flexible ties between a timber inner leaf and an outer brick leaf should be austenitic stainless steel to grade 1.4301 (to BS EN 10088-1: 2005) screwed or nailed to the timber framing members. (Copper should be to ISO 1461, and polypropylene must meet the heat, durability, elasticity and distortion requirements of BS EN 845-1 (2003) Table A.3.) When considering the type and placement of wall ties, bear in mind the requirements of sound attenuation in Building Regulations Approved Document E: 'Resistance to Sound'. The dynamic stiffness of a wall tie determines its ability to transmit vibration/sound and 'Type A' wall ties are generally recommended to keep this problem to a minimum. Type A refers to butterfly ties across 50 mm and 75 mm cavities with a minimum masonry thickness of 90 mm, but Type A can also apply to ties of appropriate measured dynamic stiffness to suit the cavity. These are the default choice of wall tie whereas 'Type B' is required for instances where Type A fails to meet the requirements of Approved Document A: 'Structural Safety'.

References

BS 4729 (2005) 'Clay and calcium silicate bricks of special shapes and sizes – Recommendations', BSI

BS 5628-2 (2005) 'Code of practice for the use of masonry. Structural use of reinforced and prestressed masonry', BSI

BS 6213 (2000) 'Guide to the selection of construction sealants', BSI.

BS 8000 (2001) 'Workmanship on building sites. Part 3: Code of practice for masonry', BSI.

BS EN 998-2 (2003) 'Specification for mortar for masonry. Part 2: Masonry mortar', BSI.

Jones, M. (2004) 'Movement Joints', Master Builder Magazine, February 2004, The Federation of Master Builders.

National House Building Council (2006) 'NHBC Standards – 6.1 External masonry walls', NHBC.

RECOMMENDED READINGS
BS 5606 (1990) 'Guide to accuracy in building' (AMD 9975), BSI.

BS 5628-1 (2005) 'Code of practice for use of masonry. Part 1: Structural use of unreinforced masonry', BSI.

BS 8000-16 (1997) 'Workmanship on building sites. Part 16: Code of practice for sealing joints in buildings using sealants', BSI.

BS EN 845-1 (2003) 'Specification for ancillary components for masonry. Part 1: Ties, tension straps, hangers and brackets' (AMD Corrigendum 14736) (AMD 15539), BSI.

BS EN ISO 11600 (2006) 'Building construction – Jointing products – Classification and requirements for sealants' (AMD Corrigendum 15975), BSI.

BRE (1974) BRE Digest 163 'Drying Out Buildings', BRE.

BRE (1991) BRE Digest 361 'Why Do Buildings Crack?' BRE.

08: Going Underground The pros and cons of basement construction

Some say that it's a cellar's market, but the basement hasn't really become a favourite in the UK construction industry. Maybe, when you're in a hole, stop digging. Basements tend to be occupied spaces under buildings (whereas cellars are usually unoccupied, storage areas) that can add volume and value to a property, especially where extending upwards is not an option.

In recent years, some private-sector and trade bodies have begun to draft Approved Documents (ADs) for the Building Regulations (E&W); an early one was TRADA's 'Timber Intermediate Floors for Dwellings' relating to AD A: Structure. One of the more recent ones being promoted on the government's Planning Portal website is the Basement Information Centre's 'Basements for Dwellings'. This AD, which hasn't been designated with a letter, is a single source pulling together all key documentation relevant to basements and more helpful guidance besides. Even though its chief sponsor was the Office of the Deputy Prime Minister (ODPM), the AD is the work of the Basement Information Centre (BIC) supported by the Concrete Centre. In addition to the AD, the BIC has produced a series of useful guides exploring the viability and costs of a basement, including advice on assessing the thermal performance, criteria for the overall design, waterproofing and remedial work.

'Basements for Dwellings' was first published in 1997 and was updated to tie in with Building Regulations 2000. At the time of writing, the current version (2004) is based on regulations in force at that time and so misses out on the recent changes in AD L, F and B. This is its drawback: as a private-sector tool it will always be playing catch-up with the regulations in force at the time of its production. While an addendum has been added to the Planning Portal, the full update is waiting on a range of prospective alterations to other ADs in 2009.

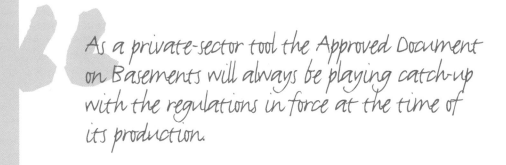

As a private-sector tool the Approved Document on Basements will always be playing catch-up with the regulations in force at the time of its production.

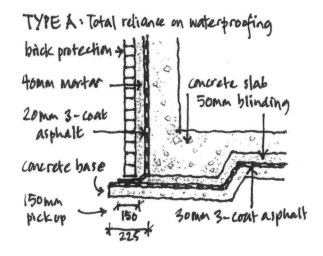

TYPE A: Total reliance on waterproofing

brick protection →

40mm mortar —

20mm 3-coat asphalt —

concrete slab 50mm blinding

concrete base

150mm pick up → 150   30mm 3-coat asphalt

225

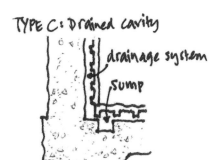

TYPE B: Water-resistant concrete

TYPE C: Drained cavity

drainage system

sump

However, it seems inevitable that, with the Building Regulations currently undergoing a review of their *raison d'être*, such private-sector packages will become increasingly influential in the way we reference information. Already 'Basements for Dwellings' is used as a guide for the Loss Prevention Council's 'Standard for Innovative Systems, Elements and Components for Residential Buildings' (LPS 2020: 2006).

Furthermore, with concerns about land use and land values, the UK government seems keen to explore ways of maximising house-building cost-benefits and addressing energy-saving considerations in the residential sector. Keith Hill MP, ex-Minister of State for Housing and Planning has said that 'basement waterproofing concerns of yesteryear are a thing of the past'. Whether this is true or not, this Shortcut pulls together some of the current legislation relating to 'Basements for Dwellings', especially regarding asphalt tanking.

A basement is defined in Scottish Technical Handbook 1: Structure as 'a storey which is below the level of the ground storey and in the case of a sloping ground level is wholly below the lowest ground level around the building'. The British Standards Institute defines it as a 'usable part of a building that is situated partly or entirely below ground level'. Even though Approved Document L: 'Conservation of Fuel and Power' relates to heated and unheated basements, 'The Code for Sustainable Homes' (which at the time of writing is a voluntary document), defines a basement as a 'heated' space, which also has 'safe access by a permanent stairway or other means of entrance which complies with the requirements of relevant national Building Regulations'. It also states that a basement should be 'finished' with floor, walls, lighting and electric sockets.

When constructing, adapting or extending a basement, Building Regulations permission is required in all cases and, if the basement is part of a workplace premises, then a Full Plans application must be deposited. The walls, ground floor and basement floor must be designed to resist the structural loads as well as imposed loads from the surrounding soil. Unless the excavations are sufficiently distant from the site boundary and adjoining buildings, The Party Wall etc. Act 1996 (see Shortcuts: Book 2) will undoubtedly be applicable. This will necessitate agreement with the adjoining owners (including leaseholders) before any work is started. Similarly, planning permission under the Town and Country Planning Act and/or the consent of the environmental health authority may also be required regardless of visibility (remember the Albert Dryden case*) and work should not start until all necessary approvals are in place. Also, detailed hazard avoidance considerations regarding, *inter alia*, the risks of working in confined spaces (if appropriate) must be spelled out at an early stage in the design process.

In the AD (making reference to BS 8102: 1990), Grade 3 basements are those designed for habitable accommodation, which means that no water penetration is permissible and normal ventilation and space heating services must ensure a 'dry environment'. This performance specification can be achieved in two ways: by externally waterproofing (or tanking) the structure (Type A) or by constructing the structure out of water-resistant materials (Type B), which may need to contain an applied moisture barrier. Neither is suitable for high or variable water table conditions, in which case Type C construction must be used. This allows controlled ingress of moisture through the outer skin and captures it in an internal wall cavity where it is channelled to a sump or discharge pipe, which, in turn, may require pumping out depending on the inverts of connecting drains. The cavity of a Type C basement can be filled with insulation in England and Wales only. Legislation in the Isle of Man and Northern Ireland does not permit full fill pumped cavity insulation to be installed during the time of construction, and Scottish legislation does not permit any full fill insulation at the time of construction.

Proprietary systems or traditional asphalt detailing are common tanking solutions (the NHBC does not allow polyethylene tanking, such as Visqueen). When using mastic asphalt it must be applied in three coats, to a thickness of 30 mm on horizontal surfaces and 20 mm on sloping and vertical surfaces. A two-coat fillet should be used to all internal angle junctions. External applications require a supporting concrete base to

HABITABLE, HEATED, NEW BUILD BASEMENTS - NOT GARAGE (TO AD B+F)

Provide alternative means of escape ADB : para 2.8

All elements to be FR30

≤5m

GL

Where the basement is more than 1.2m below the highest ground, the basement is not to be included in the calculation of number of storeys (see SHORTCUT 22: LOFTS for further implications on escape routes).

≥1.5h

(h) max 2.7m

max 2.7m

where inner room, alternative escape required

curiously this basement arrangement is to be ventilated in accordance with AD F "dwellings without basements"

Background ventilation:
61-70m² - 30,000 mm²
71-80m² - 35,000 "
81-90m² - 40,000 "
91-100m² - 45,000 "

If using continuous ventilation, use table below applied to WHOLE 'HOUSE' or 0.3 L/sec/m² whichever is greater

max open plan between structural walls 70m²

PERMANENT OPENING (classified as multi-storey building)
UP TO 2-STOREYS (ex·basement), MAX HEIGHT ≤ 5m TO TOP STOREY

Include 1100mm safety guarding to comply with AD K

as above ≥1.5h

≤5m

FR60 ↓

Max open plan area between structural walls 30m²

FR30

GL

In this arrangement continuous mech extract (with or without heat extract) is 'preferred' as cross-ventilation is difficult to achieve

VENT RATE IN BASEMENT	1B	2B	3B	4B	BEDS
	13	17	21	25	L/sec

NB: where there are no bedrooms, assume one bedroom (13 L/sec) or allow for a rate of ≥ 0.3 L/sec/m² whichever is the greater

SELF-CONTAINED BASEMENT - open on one side
MAX 2-STOREYS (ex·basement) MAX HEIGHT ≤ 5m TO TOP STOREY

Escape windows to be provided to all habitable rooms (without interconnecting doors) to be 450 x 450 mm min with 0.33m² min clear opening with lowest open area 1100mm ª FFL

≥1.5h

FR60 compartmentation Doors to be F30S no closers required

loft?

≥ 4.5m

GL

In this arrangement ventilate as a single storey building above ground. SEE NOTE ABOVE

Emergency access/egress to protected stair not passing through inner rooms

Protected stair compartment all elements to be FR 60

max open plan area between structural walls 70 m²

inner room unless protected stair + enclosure provided

NO PERMANENT OPENING (lockable/closeable separation)
TOP STOREY EXCEEDS 4.5m TO G.L.

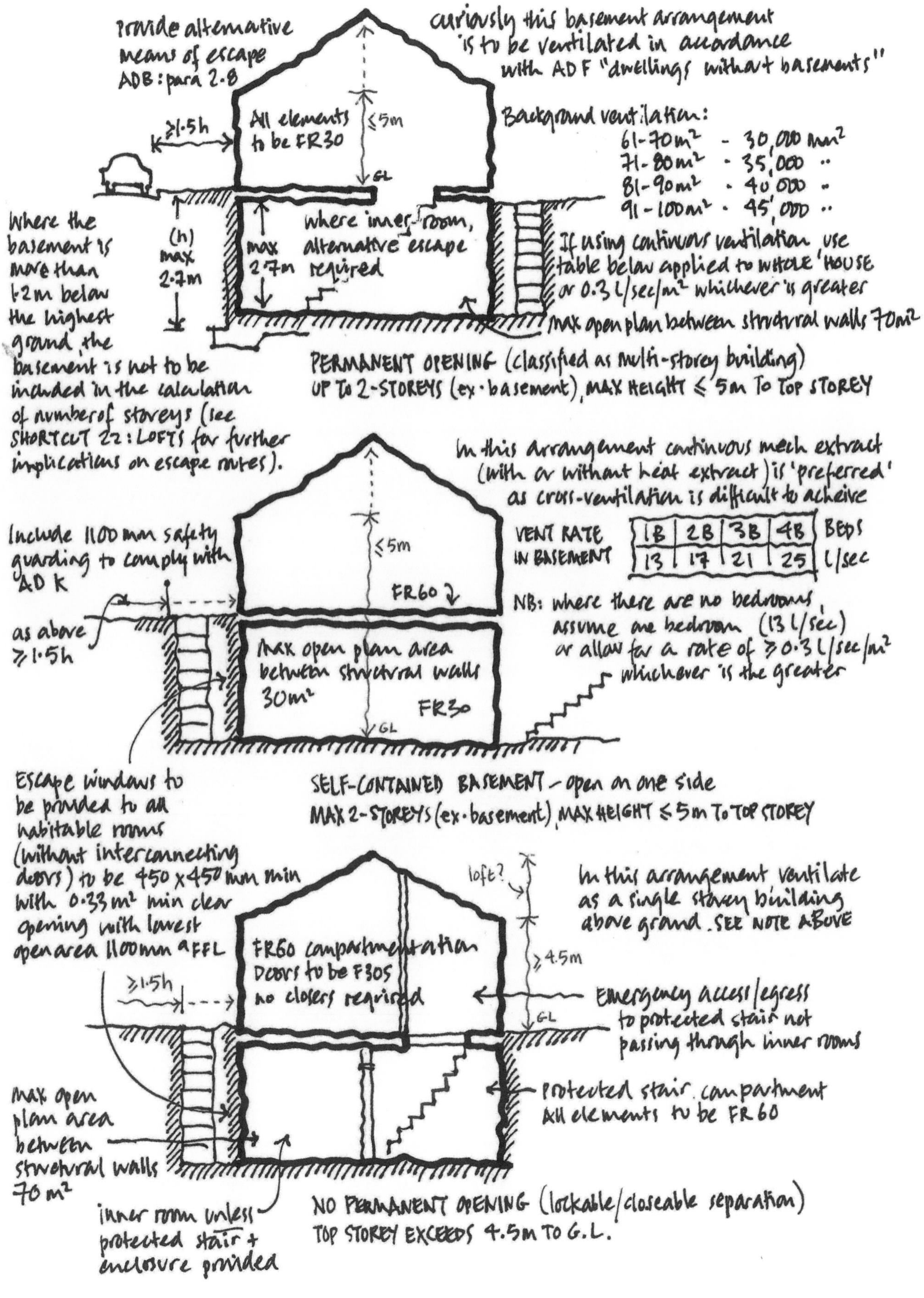

> *When constructing, adapting or extending a basement, Building Regulations permission is required in all cases and, if the basement is part of a workplace premises, then a Full Plans application must be deposited.*

extend a minimum of 150 mm beyond the outer face of the basement wall. This 'pick-up' strip, together with an angle fillet, facilitates full fusion between horizontal and vertical coats. The BRE states that mastic tanking will provide sufficient radon protection to make supplementary protection (e.g. a sump) unnecessary.

Upon completion of the horizontal tanking to the concrete base, a protective cement:sand screed of 50 mm minimum thickness is immediately laid to protect against following trades, and a loading coat of concrete, designed to resist the maximum anticipated water pressure, should follow. The 'pick-up' strip of asphalt must not be contaminated, and the screed, which offers only temporary protection, is laid on building paper local to the strip, to facilitate removal. This allows construction of a brick or block wall to protect the vertical membrane from the effects of solar heating and potential damage during the course of backfilling operations. This wall should be erected to provide a 40 mm gap from the mastic asphalt membrane, which should be filled with mortar, course by course, as the work progresses. In some instances, rot-free boards can be used to provide vertical protection.

In terms of central heating requirements, LPG boilers must not be used in basements unless there are specifically sanctioned for the particulars of the site. Gas boilers are acceptable, although provision must be made for air supply and the disposal of condensate (a condensate pump might be required if a drain point cannot be reached by gravity). The size of oil-fired boilers and storage tanks tends to make them more difficult to incorporate within basements. Pump manufacturers' instructions must always be followed and the disposal of the products of combustion must be dispersed in accordance with specialist (Corgi and OFTEC) organisation recommendations; flue terminals, for example, must be at least 300 mm above external ground level. Balanced flue terminals must not be located in light wells, etc., but taken to suitably ventilated external spaces.

* In 1991, Albert Dryden from Consett wanted to build a house but was in dispute with Derwentside District Council planning authority. To avoid detection, he dug a huge hole and built his traditional bungalow surrounded by banked up ground. When the planners arrived with TV crews to serve notice of demolition, Dryden shot chief planning officer, Harry Collinson, dead.

GUIDE TO THE VARIOUS LEVELS OF PROTECTION (FROM BS 8102: 1990)

Grade	Use	Watertightness	Type of construction
1	Car park, plant, workshops	No water penetration, some damp patches	Type A or Type B (reinforced concrete to BS 8110)
2	Plant, workshops, retail storage	No water penetration, tolerable moisture vapour	Type A or Type B (reinforced concrete to BS 8007)
3	Ventilated residential, offices, etc.	Dry environment	Type A, Type B (reinforced concrete to BS 8007) or Type C (with wall and floor cavity and dpm)
4	Archives, controlled storage	Totally dry	Type A, Type B (reinforced concrete to BS 8007 with vapour proof layer) or Type C (with ventilated wall cavity with vapour proof membrane to inner skin and with floor cavity and dpm)

References

Anderson, B. (2006) BR 443 'Conventions for U-value calculations', BRE.

Blackie, D. (2006) 'Death on a Summer's Day', John Blake Publishing.

British Cement Association (1994) 'Basement waterproofing – design guide', BCA.

BS 6100-0 (2002) 'Glossary of building and civil engineering terms: Part 0: Introduction', BSI.

BS 8102 (1990) 'Code of practice for protection of structures against water from the ground', BSI.

Building Research Establishment (1999) Good Repair Guide 23 'Treating dampness in basements', BRE.

Steel Construction Institute (2001) Publication 275 'Steel intensive basements', SCI.

RECOMMENDED READINGS

Building Research Establishment (2007) Report 487, Chapter 4 'Designing Quality Buildings: A BRE guide', BRE.

Building Research Establishment (2002) Report 440 'BRE building elements: foundations, basements and external works', BRE.

National House Builders' Confederation (2006) Chapter 5.1 'Substructure and ground bearing floors', NHBC.

The Basement Information Centre (2004) 'Approved Document – Basements for Dwellings', TBIC.

The Basement Information Centre guidance. Basements 1–4. Available at: http://tbic.org.uk/

09: A Course in Mortar … and when to use the 'f' in sulfur

When specifying mortar mixes, the ratio of their constituent elements should relate to their mass, rather than their volume, in order to provide batch consistency. This paper sets out some of the mix requirements for various locations using factory-produced, ready-to-use mortars.

In 1990, the International Union of Pure and Applied Chemistry recommended that the official spelling of 'sulphur' become 'sulfur' to suit the American audience (to be fair, the US 'aluminum' was recommended to change to 'aluminium'). Ten years later, and a UK 'sulfur'-resisting lobby managed successfully to challenge this mandate and, in schools at least, sulphur has returned to its high pH content.

The 'f' spelling originated in Webster's English Dictionary of 1806, which went on to became the standard simplified American version of over-complicated spellings. Noah Webster intended that all extraneous letters be removed from words for ease of understanding as well as a way of providing a new identity for the emerging USA. Apparently, a slightly conservative publisher encouraged him to back down on spelling 'tough' (tuf), 'group' (groop) and tongue (tung) but plenty of other words became Americanized. In this Shortcut, we'll stick to the 'f' spelling – accepted by the Oxford English Dictionary.

The 'f' spelling originated in Noah Webster's English Dictionary of 1806, which went on to became the standard simplified American version of over-complicated spellings.

KEY

clay brick → ← calcium silicate

concrete brick → ← concrete block

MORTAR GROUPS and suffix

KEY	MORTAR GROUPS AS BS 5628-3 table 13
SRC	Sulfate-resisting cement
a	Ensure that wall is fully protected from freeze-thaw, if not, use Group 3
b	Use SRC if normal soluble salt content clay bricks specified
c	Use SRC if normal soluble salt content clay bricks specified, if not, use Group 3
d	As b, but also increase strength to Group 2 in this example
f	Use SRC for chimney capping / coping

*NOTE: NBS uses Arabic numerals (1,2,3,4) for the Roman numerals (i,ii,iii,iv) mortar groups in BS5628 "Code of Practice for the Use of Masonry".

In many instances, a mortar mix with a cement:sand ratio of 1:3 will be strong for most uses. However, there are a variety of locations where mortar constituents and resultant strength is key to the integrity of the structure. While it might seem easier to specify one mortar designation and stick with it in all locations, the implications would be serious. Flexibility, durability and moisture- and corrosion-resistance are all factors specific to site conditions that need to be taken into account. Hardly surprisingly, sulfur-resisting cement (SRC) is used in mortars in conditions where sulfurous attack is likely to be damaging. Whereas many designers are aware that this relates to certain soils which have a high acid content liable to cause groundwater-borne sulfates, it can also relate to areas of high pollution (city centres generally, or around chimneys or gas flues, etc.).

The drawings in this Shortcut give the group references for different types of masonry and locations. The various mixes that these group numbers represent are contained in four 'designations' in BS:5628-3: 2005 Tables 12 and 13. (Note: In a peculiarity of the British Standards Institute, the superseded 2001 version is still cited in the Building Regulations, but the current 2005 edition goes into more detail.) While the mortar

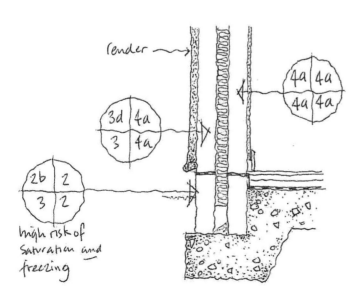

TYPICAL WALL CONSTRUCTION

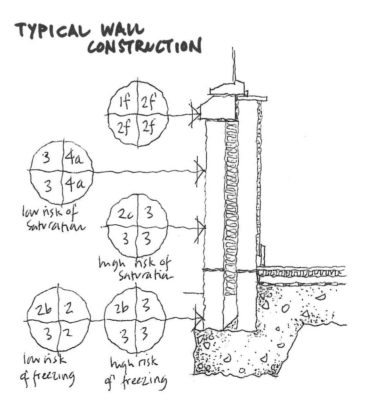

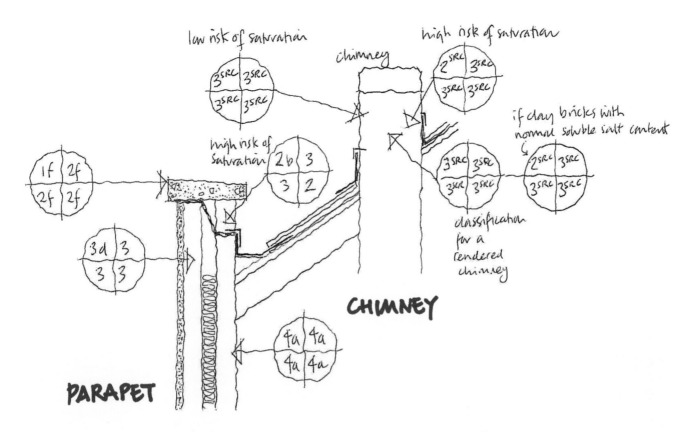

low risk of saturation

high risk of saturation

chimney

3SRC	3SRC
3SRC	3SRC

2	3SRC
3SRC	3SRC

if clay bricks with normal soluble salt content

high risk of saturation

2b	3
3	2

1f	2f
2f	2f

3SRC	3SRC
3SRC	3SRC

2SRC	3SRC
3SRC	3SRC

classification for a rendered chimney

3d	3
3	3

CHIMNEY

4a	4a
4a	4a

PARAPET

NOTE:

When using mortar for the refurbishment/ conservation of historic building projects, it is usual to use a mortar that is compatable with the original.

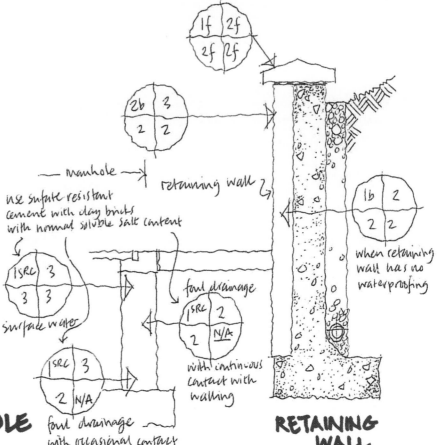

1f	2f
2f	2f

2b	3
2	2

— manhole —

retaining wall

use sulfate resistant cement with clay bricks with normal soluble salt content

1b	2
2	2

1SRC	3
3	3

foul drainage

when retaining wall has no waterproofing

surface water

1SRC	2
2	N/A

with continuous contact with walling

1SRC	3
2	N/A

MANHOLE

foul drainage with occasional contact with walling

RETAINING WALL

designation groups are written in this Shortcut and in NBS guidance as Groups 1 to 4 (inclusive), they can also be written as Groups i to iv (appearing as Roman numerals in the British Standard). Specifying to these mortar groups will suffice for specification purposes, as the BS identifies the actual mix criteria. However, you will have to specify whether you want the mix to include lime, otherwise the mix references refer to a simple sand:cement mix.

The lower the number, the higher the compressive strength. For example, Group 1 (i) is around 12 N/mm^2 at 28 days, while Group 4 (iv) is just 2 N/mm^2. However, it is an exponential change given that Group 2 (ii), is 6 N/mm^2 and Group 3 (iii) is 4 N/mm^2.

The higher compressive strength groups provide durability and structural stability and are usually recommended for external, weathered or below-ground structural work, while the low compressive strength groups have greater flexibility and are usually suitable for internal, protected masonry and areas prone to thermal movement, including chimneys.

> "Water used for mixing the mortar should be free of contaminants. A general indicator is that if it is suitable for drinking, it is suitable for mortar mixing.

References

BS 4551 (2005) *Methods of testing mortars, screeds and plasters'*, BSI.

BS 5628-3 (2005) *'Code of Practice for the use of masonry – Part 3: Materials and components design and workmanship'*, BSI.

BS EN 197-1 (2000) *'Cement composition, specification and conformity criteria for common cements (AMD 15209) (AMD 17352)'*, BSI.

BS EN 459-1 (2001) *Building lime. Definitions, specifications and conformity criteria (AMD Corrigendum 14099)*, BSI.

BS EN 771-1 (2003) *'Specification for masonry units – Part 1: Clay masonry units (AMD 15998)'*, BSI.

BS EN 772-3 (1998) *'Methods of test for masonry units – Part 3: Determination of net volume and percentage of voids of clay masonry units by hydrostatic weighing'*, BSI.

BS EN 772-7 (1998) *'Methods of test for masonry units – Part 7: Determination of water absorption of clay masonry damp proof course units by boiling in water'*, BSI.

BS EN 1015 (1999) *'Methods of test for mortar for masonry. Determination of flexural and compressive strength of hardened mortar (AMD 16880)'*, BSI.

BS EN 13139 (2002) *'Aggregates for mortar (AMD Corrigendum 15335)'*. BSI.

Building Research Establishment (2001) *'BRE Special Digest 1: Concrete in Aggressive Ground'*, BRE.

RECOMMENDED READINGS

BS 998-2 (2002) *'Specification for mortar for masonry – Part 2: Masonry mortar'*, BSI.

BS 5628 (2005) *'Code of practice for use of masonry'*, BSI.

PD 6678 (2005) *'Guide to the selection and specification of masonry mortar'*, BSI.

PD 6682-3 (2003) *'Aggregates for mortar – Guidance on the use of BS EN 13139'*, BSI.

10: Soil Nailing Stabilising sloping ground

As far as most architects are concerned, the angle of repose refers either to the maximum slope that soil can be banked before it starts to slide under its own weight; or it is the optimum ergonomically designed position of the leather recliner on a Friday afternoon. This Shortcut looks at the former.

Soil nailing is a way of reinforcing sloping ground conditions by the insertion of tension-carrying elements called soil nails. Essentially, the material in question, say for example, a sloping bank of natural ground or fill is pinned back using 'nails' to prevent it moving. A soil-nailed slope usually consists of the soil nails themselves (typically a series of 20–30 mm diameter threaded coated steel rods), a hard, flexible or soft facing to the slope surface, and surface water and sub-surface drainage systems.

By using soil nails, angled ground can be stabilised at a much greater pitch than would otherwise be the case. In a highways situation, for example, for a given road width cut into a bank, the excavated soil need not be taken back at as shallow a rake as would be required if the natural stability of the soil were being relied upon, thus necessitating less material to be removed. In some instances, soil nailing can be used in lieu of a retaining wall or can be added to a retaining wall specification to improve its efficiency. It can also be used as a remedial treatment on existing retaining structures.

The Construction Industry Research and Information Association (CIRIA) has produced the definitive guide, 'Soil nailing – best practice guidance', which notes that the system has only been tried out in the UK since the mid 1980s. By the mid 1990s, only 3000 linear metres of soil-nailed surfaces had been created (whereas, in France, 30 times as much was being created every year at that time). For those first 5–10 years, British practice relied solely on the evidence of the practical work experienced in mainland Europe, and as such there were no guidelines, technical regulations or substantive scientific research to justify the suitability of the procedures in the UK.

How it works The principle is straightforward – it is the engineering for the site-specific requirements that is the complex bit. In terms of stabilising natural slopes alongside railway embankments for instance, initial tests are carried out to ascertain the density, cohesiveness, fluidity, etc. of the soil. This will help to determine the number, angle, size, type and location of the nails. Soil nailing should not be undertaken without a detailed geotechnical analysis of empirical data (soil nailing is not generally recommended for soft clay soils, for example – those with an undrained shear strength of 48 kN/m^2).[1]

The main engineering concern in the design of these retaining systems is to ensure that the specification of the soil nails suits the conditions, in order to adequately secure and stabilise the ground with an appropriate factor of safety. It is important to note that the

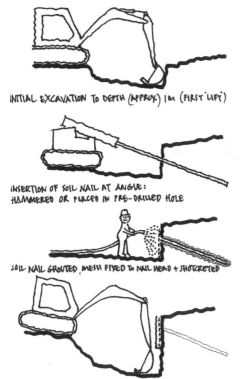

INITIAL EXCAVATION TO DEPTH (APPROX) 1M (FIRST LIFT)

INSERTION OF SOIL NAIL AT ANGLE:
HAMMERED OR PLACED IN PRE-DRILLED HOLE

SOIL NAIL GROUTED, MESH FIXED TO NAIL HEAD + SHOTCRETED

CAREFUL SECOND LIFT AND PROCESS REPEATED

compression – and general ground movement – caused by the insertion of the nails can potential cause damage to any existing drainage or other services in the vicinity, and this must be taken into consideration when assessing whether the method is appropriate.

Once the pattern and specification of the nails is confirmed, a mobile jig is usually required to drive, drill or hammer the nails into the soil. Note: Sometimes, the operational width available on tight sites (say, on narrow railway cuttings as opposed to railway embankments) will be insufficient for mobile drilling equipment and therefore hand-held hammer drills or crane jib-mounted drills may be necessary, with the consequent increased health and safety risk factors.

As the soil nails are punched into the ground at an angle, they transfer tensile forces into the substrate via the frictional resistance between the nails and the surrounding soil. Those that are punched into embankments are usually also grouted along their length to provide maximum frictional pull-out resistance and to ensure that these friction forces optimise the overall shear resistance of the native ground – effectively creating a much larger zone of resistance against displacement. Grouting is often carried out simultaneously with the drilling and the soil nail insertion process to ensure a good bond. In some drilling systems, the grout is circulated within the borehole as the nail is being inserted in order to mix the grout with the drill spoil to form a strong 'grout bulb'. A pressure release hole at the top helps to flush out some of the lighter drill cuttings.

Soil nailing not only works in tension, but also induces bending and shearing forces, and the nail size and material must take this into account. The nails tend not to be used in isolation but are attached to a head pad which increases the area of the nail head, providing additional load spread and an increased area of soil retention. In many instances, a flexible net or a rigid concrete surface can be incorporated between nail heads to help retain the ground surface layer. Nets tend to be used on 'shallower' slopes where planting might be cultivated. In some locations, degradable coir matting is being used (in lieu of the ubiquitous geogrid materials) where it has been assessed that, as the planting becomes established, it alone will provide sufficient cohesion to the slope. On steeply inclined, heavily loaded ground, or in areas that need to resist impact and/or to have a different aesthetic appeal, a sprayed concrete (shotcrete) facing – reinforced using woven mesh – is usually applied. In vertical instances, permanent walls can be built with prefabricated or in-situ panels affixed to the nail head plates.

Background The practice of soil nailing originated in France in the early 1960s in order to stabilise rock excavations in the construction of mountain passes. The first soil-nailed wall was built at Versailles in 1972/3. The technique was further developed with the introduction of a revolutionary tunnelling method devised in Austria in the late 1960s. Unsurprisingly known as the (New) Austrian Tunnelling Method (NATM) – this technique minimised excavations and eliminated the need for reinforced concrete structural formwork. Engineering academics Karaku and Fowell insist that the NATM method of soil nailing is not an actual 'technique' but is simply a 'philosophical approach to excavation'. Whatever the resolution to that argument, it is indisputable that it has become a well-accepted approach to tunnel building, and works by using soil nails, together with a thin reinforced concrete lining, to maximise the self-supporting capacity of the excavated rock or soil. The nailing process coheres blocks of earth or substrate, and a grid of nails is thereby able to hold back the tunnel overburden.

Following the collapse of the Heathrow Express Rail Link Station on 21 October 1994, a number of questions have been raised about the efficacy of this tunnelling method. However, the Transport Research Laboratory (TRL) was instrumental in alleviating concerns over the durability of the system, and soil nailing guides were developed for stabilising highway embankments. However, given the almost non-existent level of road-building in this country in the past decade, most of the UK market deals with retaining existing ground rather than constructing new.

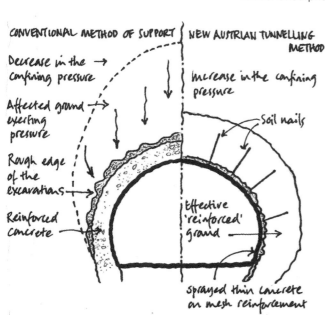

CONVENTIONAL METHOD OF SUPPORT | NEW AUSTRIAN TUNNELLING METHOD

Decrease in the → confining pressure

Affected ground → exerting pressure

Rough edge of the excavations →

Reinforced concrete

Increase in the confining pressure

Soil nails

effective 'reinforced' ground

sprayed thin concrete on mesh reinforcement

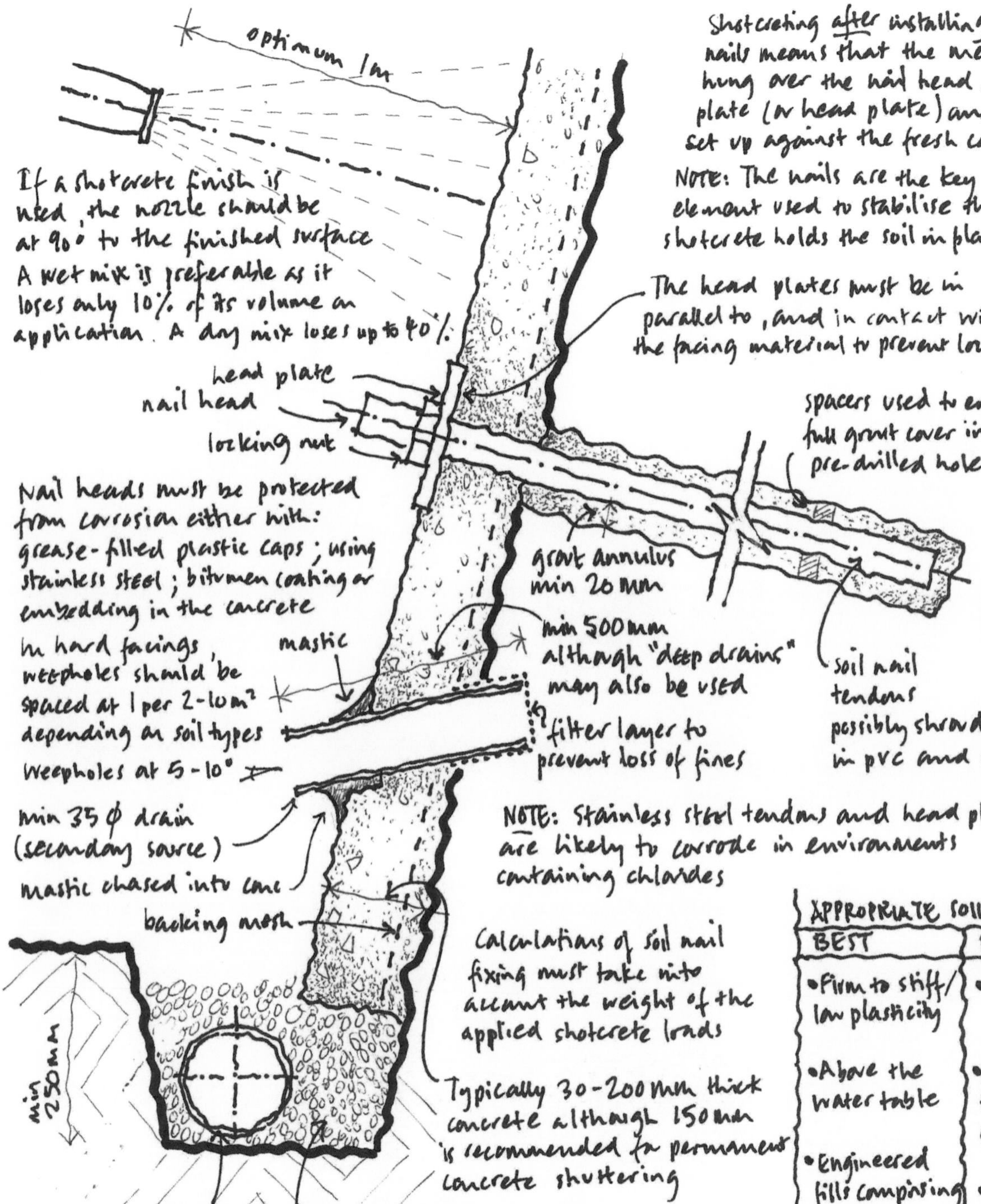

optimum 1m

If a shotcrete finish is used, the nozzle should be at 90° to the finished surface. A wet mix is preferable as it loses only 10% of its volume on application. A dry mix loses up to 40%.

head plate
nail head
locking nut

Nail heads must be protected from corrosion either with: grease-filled plastic caps; using stainless steel; bitumen coating or embedding in the concrete

In hard facings, weepholes should be spaced at 1 per 2-10m² depending on soil types. Weepholes at 5-10°

mastic

min 35 ⌀ drain (secondary source)
mastic chased into conc
backing mesh

min 250mm

100-150mm French drain (the primary source of water management)

Shotcreting after installing the soil nails means that the mesh can be hung over the nail head and bearing plate (or head plate) and can be set up against the fresh concrete

NOTE: The nails are the key structural element used to stabilise the soil. The shotcrete holds the soil in place between nails

The head plates must be in parallel to, and in contact with the facing material to prevent local stresses

spacers used to ensure full grout cover in pre-drilled holes

grout annulus min 20mm

min 500mm although "deep drains" may also be used

filter layer to prevent loss of fines

soil nail tendons possibly shrouded in PVC and grouted in

NOTE: Stainless steel tendons and head plates are likely to corrode in environments containing chlorides

Calculations of soil nail fixing must take into account the weight of the applied shotcrete loads

Typically 30-200mm thick concrete although 150mm is recommended for permanent concrete shuttering

compacted fill or no-fines concrete

FINISHED RETAINING SURFACE:
· SOFT FACING:
Soil nails and plastic coated 80x80mm mesh with 'vertical planting' (or geotextile impregnated with grass seeding)
· HARD FACING:
Spray concrete on mesh backing (shown) or prefabricated units attached to the head plates

APPROPRIATE SOILS FOR NAILING	
BEST	POSSIBLE
•Firm to stiff/low plasticity	•Soft cohesive
•Above the water table	•At the water table in granular soils
•Engineered fills comprising natural non-aggressive materials	•Non-engineered fills
	•Loose sands
	•Degraded materials
•Medium to dense sands and gravel with some cohesion	•Frost susceptible soils and rocks

> *In new excavations, as opposed to remedial work to existing excavations, nails should be inserted in stages, as the excavations proceed. Because of the various pressures acting on the excavations, the nails must be able to resist tensile and shear stresses as well as bending moments.*

Soil nailing is an economical means of creating shoring systems and retaining walls and nailing is often less disruptive and cheaper than other means of constructing retaining systems.

A recent Department of Transport survey revealed that, of the '5400 km of walls supporting highways in the UK', 50 per cent are dry stone! The need to provide heightened assurances that these structures are safe from landslides is one of the driving forces for pushing soil nailing research and development.

Staged dig Factors to consider in initial soil nailing analysis are:

- the water table
- the shrinkage (and the propensity to shrink) of the retained soil layers
- any underlying geological traits
- the need for adequate groundwater drainage.

Poor drainage may result in the swelling of clays behind the facing layer resulting in localised collapse; otherwise the build-up of water (possibly carrying with it sulfates or other aggressive chemicals) behind the main retained area may simply erode the bond between the nail and the ground, or between the sprayed concrete and the substrate, leading to significant leakage and potential collapse.

In new excavations, as opposed to remedial work to existing excavations, nails should be inserted in stages, as the excavations proceed. Because of the various pressures acting on the excavations, the nails must be able to resist tensile and shear stresses as well as bending moments. They ought to be double corrosion protected too, in accordance with BS 8081, 'Code of practice for ground anchorages' (this standard is partially superseded BS EN 1537: 2000 but remains current). This tends to mean a standard protective coated steel nail pre-grouted within a plastic-coated outer shell. The plastic shell is heavily corrugated for extra grip. (Note: In express contradiction to BS 8081, Network Rail's soil nail specification states that the grout – together with a sacrificial nail thickness – is considered to be suitable 'double protection'.

Where excavations requiring soil nailing are being carried out, the ground should be reduced by about 1–2 m in height and the first nail inserted. (Note: The soil must have adequate 'apparent cohesion' to enable the dig to be carried out in the first place, without the need for temporary retaining formwork. Typically, after the nails are inserted and the head plates attached, welded mesh will be affixed to nail heads and a 100–200 mm thick sprayed concrete facing applied. Once this is done and set, the next cut can be made and the procedure repeated.)

Depending on the type of scheme, a drainage trench should be incorporated at the top (or 'crest') of the slope or wall, taking rainwater run-off, with weepholes along the face of the slope or wall (possibly in conjunction with horizontal French drains) and a 'toe drain' at the base. Also, allow for interceptors and attenuators to manage the flow.

[1] Gladstone Bell, F. (1993), *'Engineering Treatment of Soils'*, Taylor & Francis, p.167

References

Bruce, D.A. & Jewell, R.A. (1986) *'Soil nailing: application and practice – part 1'*, Ground Engineering, Nov, pp. 10–15.

BS 1377 (1990) *'British standard methods of test for soils for civil engineering purposes'*, BSI.

Building Research Establishment (1995) *'Sulphate and acid attack on concrete in the ground: recommended procedures for soil analysis'*, BR 279, BRE.

Charles, J.A. & Watts, K.S. (2001) *'Building on fill: geotechnical aspects'*, Report 424, Building Research Establishment, BRE.

Coppin, N. & Richards, I. (1990) *'Use of vegetation in civil engineering'*, Book 10, CIRIA, London and Butterworth.

Driscoll, R. & Simpson, B. (2001) *'EN 1997 Eurocode 7: Geotechnical design'*, Proc Instn Civ Engrs, Civil Engineering Journal, vol. 144, Nov, pp. 49–54.

Terzaghi K. & Peck, R.B. (1948 & 1967) *'Soil Mechanics in Engineering Practice'*, John Wiley and Sons.

RECOMMENDED READINGS

British Drilling Association (2002) *'BDA health and safety manual. A code of safe drilling practice'*, BDA.

BS 8081 (1989) *'Code of practice for ground anchorages'*, BSI.

BS EN 1537 (2000) *'Execution of special geotechnical work, ground anchors'*, BSI.

BS EN 1997-1 (2004) *'Eurocode 7. Geotechnical design. General rules'*, BSI.

Johnson, P.E., Card, G.B. & Darley, P. (2002) *'Soil nailing for slopes'*, Report 537, Transport Research Laboratory.

Phear, A., Dew, C., Ozsoy, B., Wharmby, N.J., Judge, J. & Barley, A.D. (2005) *'Soil nailing – best practice guidance'*, CIRIA C637.

PrEN 14490 (2002) *'Execution of special geotechnical works – soil nailing'*, (draft standard).

11: Flat Roofing
Getting your collar felt

Ensuring that a flat roof has the correct slope is an essential first step to low maintenance. The second step is accommodating movement. Here we explore some of the key aspects of flat roofs, junctions and abutments, and look at some general methods of ensuring rainwater run-off.

Everyone is health and safety conscious these days, with concerns about personal safety paramount in the construction industry. But sometimes, avoiding falls at all costs is the wrong thing to do. After all, when it comes to flat roofs, the more falls, the better.

Admittedly, gone are the days when builders laid joists with the aid of a spirit level, or argued that a firring was a problem for a plumber. We all know that a flat roof isn't meant to be truly flat. Hopefully, enough flat roofs have been commissioned since the leaky days of the 1970s, and enough advances have been made in the materials available, for the lessons of the past to be learned. Today, there is a much more credible understanding of the potential problems and their solutions. However, there are still errors being made on sites – large and small – across the country, and so here we provide a snapshot of the crucial factors relating to falls. Other issues of roofing and detailing will be dealt with in later Shortcuts.

FALLING WATER
A flat roof is defined in BS 6229 as one with a fall of no more than 10°, while roofs between 10°–22° are often known as sloping roofs. Any more than this and the phrase 'pitched roof' generally applies, except that any roof sloping more than 70° is considered to be a wall (in terms of its acoustic and insulation performance standards). For example, the Scottish Technical Handbook, Section 1: 'Domestic' states that: 'In calculating the number of storeys for the purpose of determining if a building has five or more storeys, no account shall be taken of any storey within a roof space where the slope of the roof does not exceed 70 degrees to the horizontal.'

In essence, failure of external insulation system roofs should only occur as a result of contractor error. In reality, poor detailing frequently manifests itself many months after installation.

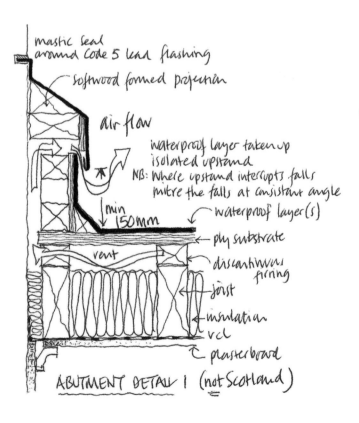

The guidance in BS 6229 advises designers on the minimum falls required for a variety of materials. These should be the falls anticipated on the completed, fully-loaded roof. Roofs – especially timber and metal deck roofs – flex under snow and live loads, for example, and the consequential deflection (flattening out) of the slope can lead to unanticipated failure. In the worst case scenario, undrainable snow loads (or snow that is not removed in some other way) can lead to collapse. This is reputed to be the reason for the Moscow market roof collapse in February 2006, a tragedy that killed 49 people.

The current version of BS 6229 (2003) is only the second edition of this key statutory instrument (the first edition having been published way back in 1982 when single ply roofing was just a glint in a British architect's eye). In the first edition's version of this table, in lieu of 'single ply', there was that architectural favourite, the 'semi-rigid asbestos bitumen sheet'. Ah, the good old days.

INSTALLATION GUIDELINES

While it is well known that bitumen-based products, and flat roofing generally, received a hail of protest from the mid 1970s onwards, let's not forget that the higher technology single plies, such as EPDM (ethyl polymer diene monomer), have sometimes been prone to failure, usually at the glued joints. It is only in the past 15 years that thermoplastic polyolefins (TPOs) have been satisfactorily tested in roofing conditions down to a minimum 1:80 fall. Perhaps it is because most technical/practical guidelines were written in the early 1990s that many manufacturers still recommend minimum falls be designed to 1:40, just to be on the safe side.

For all roof constructions, it is beneficial for the waterproofing layer (or system) to be isolated from the substrate, to protect it from any differential movement that might otherwise cause it to split. This can be done either by partially bonding the roofing material, or preferably by laying an isolating layer before the waterproof layer (or system) is fixed. This will allow the substrate to move without necessarily tugging on the waterproofing. At designed movement joints, abutments, etc., it is essential that the waterproofing be discontinued without compromising either its integrity, or the ventilation or the flow of the roof falls. Any upstands that block the flow of the falls will need additional rainwater outlets to prevent ponding. Routine maintenance/regular inspections for roof membranes may also be necessary to reduce the potential for failure.

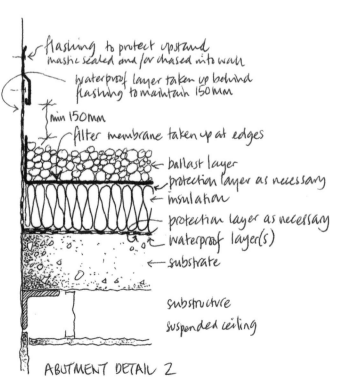

TPOs and other single-ply materials have been used under green roofs (see Shortcuts: Book 2) and externally insulated roofs for some time. Such external insulation systems (often referred to as inverted roofs because they reverse the traditional placing of insulation below the substrate and waterproof layer) are made up of the roof substrate overlaid with the membrane, with the insulation placed above that. Usually, the insulation is ballasted to prevent wind uplift. The idea is that condensation will form outside the building structure thus eliminating any potential moisture damage internally. However, there must be suitable protective layers to prevent mechanical damage during installation and maintenance, as well as suitable paths underneath the system to allow water run-off.

Theoretically, condensation occurs within the thickness of the insulation. In practice, though, it tends to form at the interstice between the waterproofing membrane and the insulation.

WATERPROOF INTEGRITY

Problems with inverted and green roofs arise when leaks appear. Leaks, which are notoriously difficult to trace in 'normal' flat roofs are doubly troubling for owners of inverted roofs, as they may require large areas of ballast to be removed to locate the failure point. However, good falls, good drainage and good detailing can minimise the potential for nominal failure to cause internal damage. The idea is that failure only occurs as a result of construction problems – as opposed to the mechanical damage, climate impact and material wear, that affect normal flat roof systems – because the membrane is protected beneath the insulation and ballast.

Where surface drainage systems are included, ensure that there is a smooth transition into the outlet. Any 'lip' between the roof fall and the outlet will cause ponding and possible capillary action through any poorly sealed joints.

Where upstands are necessary at abutments, the waterproofing must be taken a minimum of 150 mm above the uppermost roofing surface. For example, in Detail 1, the upstand from the roofing surface to the bottom-most point of the copper-clipped lead flashing must be 150 mm at all times. In Detail 2, the 150 mm upstand distance must be measured from the top of the ballast layer. Care must be taken to ensure that there is no reduction of this distance caused by the ballast shifting during high winds (known as 'wind scour') or piling up during maintenance to reduce that distance.

> *It is only in the last 15 years or so that thermoplastic polyolefins (TPOs) have been satisfactorily tested in roofing conditions down to a minimum 1:80 fall.*

COVERING	MINIMUM FALLS at any point	
Aluminium	1:60	(0.95°)
Copper	1:60	(0.95°)
Zinc	1:60	(0.95°)
Lead sheet	1:80	(0.58°)
Built up bitumen sheet	1:80	(0.58°)
Mastic asphalt	1:80	(0.58°)
Single ply membranes	1:80	(0.58°)
Liquid waterproofing systems*	1:80	(0.58°)

NOTE: These recommendations are applicable both to the general area of the roof & to any formed internal gutters

*For certain specialist systems designed solely for buried applications, such as garden roofs, podia and some car parks, specific reference should be made to the manufacturer's documented advice and to British Board of Agrément (BBA) certification

FLAT ROOF COVERINGS EXPLAINED:

Built-up membranes – Layers of roof covering, usually bituminous impregnated rolls laid in several staggered layers.

Single ply – A single sheet waterproof layer. It usually refers to polymeric materials.

Polymeric – Man-made organic compounds formed by combining long chains of component molecules.

Thermoplastic – This group includes PVC (polyvinyl chloride), CPE (chlorinated polyethylene), CSM (chlorosulfonated polyethylene), VET (vinyl ethylene terpolymer) and PIB (polyisobutylene). Polymeric materials soften when heated and become brittle at low temperatures, although this is reversible. These materials are incompatible with organic solvents.

Elastomeric – Includes EPDM (ethylene propylene diene terpolymer [monomer]) and butyl rubber. These are more elastic polymeric materials that are more resistant to chemicals and solvents and therefore more resistant to thermal shock with better thermal recovery.

PVC – See 'Thermoplastic'. Not compatible with bituminous materials. Thermal- or solvent-welded.

CPE – High-density polyethylene containing chloride. Flexible without the need for a plasticiser, is flame-resistant and is compatible with bituminous products. Thermal- or solvent-welded.

CSM – See 'Thermoplastic'. Highly resistant to chemical attack but difficult to seal and joint. Doesn't fade quickly.

VET – Modified to increase flexibility and can be heat- or solvent-welded. Compatible with bitumen.

PIB – Modified by the addition of carbon black to improve physical properties. Suitable for solvent welding by tape systems.

EPDM – A copolymer of ethylene and propylene with an unsaturated diene monomer. Can be vulcanised and has good weathering capabilities. Resistant to UV and ozone.

Butyl rubber – This is a copolymer of isobutylene with isoprene or butadiene. Good weathering and durability.

Flexible polyolefin – Polypropylene or polyethylene combined with flexible comonomers. Thicker and stiffer than conventional single-ply sheets. It weathers well and has excellent puncture resistance, chemical compatibility and flexibility.

Adhesive fixed – Fully adhering a waterproof layer to the substrate or bonding a layer to the substrate via a loose-laid perforated layer or by spot bonding (partial bonding). In fully or partially adhered systems, the membrane and adhesive must be compatible with the insulation in warm roof conditions. Polystyrene, for example, will melt if overlaid with hot-bonded bituminous felt.

Loose-laid – Common in ballasted single-ply systems; the edge wind uplift must be catered for by restraints and fixings.

Mechanically fastened – Edges of membranes fixed by screws and fixing plates which are bonded to the roofing material to ensure no breech of the roofing material's integrity.

HEALTH AND SAFETY

The Department of Trade and Industry guidance on flat-roof health and safety recommends that hazards be removed and residual risks identified. In some respects, flat roofs could be argued to have removed many of the slip hazards of pitched roofs, owing to their shallower pitch. More usefully, there are a number of points that should be considered as part of a considerate health and safety design risk assessment procedure for flat roofs. For example, plant should not be located at roof edges; access doors should open inwards (to avoid the potential for wind to blow them shut, or to blow open and dislodge a person on the roof); and glazed openings located within 2.5 metres from the roof edge should be of laminated materials.

It also recommends that hours of working be limited – 'especially during hot weather' – so architects may have to consider a broader interpretation of the traditional 'adverse weather' (e.g. strong winds over 10 m/sec) contingency.

References

British Flat Roofing Council/Construction Industry Research and Information Association (1993) 'Flat roofing: design and good practice', CIRIA.

BS 6229 (2003) 'Flat roofs with continuously supported coverings', BSI.

Construction Industry Council (2004) Technical guidance note T 20.009: 'Designing to make management of hazards associated with working on roofs easier', CDM. http://www.safetyindesign.org

Flat Roofing Alliance (1993) Roofing Handbook Information Sheet 21: 'Upstands and Skirtings', FRA.

Flat Roofing Alliance (1993) Roofing Handbook Information Sheet 27: 'Improving falls on Existing Roofs', FRA.

National House Building Council (2008) 'NHBC Standards 2008. Part 7 – Roofs' (Chapter 7.1 Flat roofs and balconies), NHBC.

Scottish Building Standards Agency (2006) 'Scottish Technical Handbooks', SBSA.

Single Ply Roofing Association (2003) 'Design Guide for single ply roofing', SPRA.

The Stationery Office 'The Work at Height Regulations 2005', TSO.

Part 2
FABRIC AND FINISHES

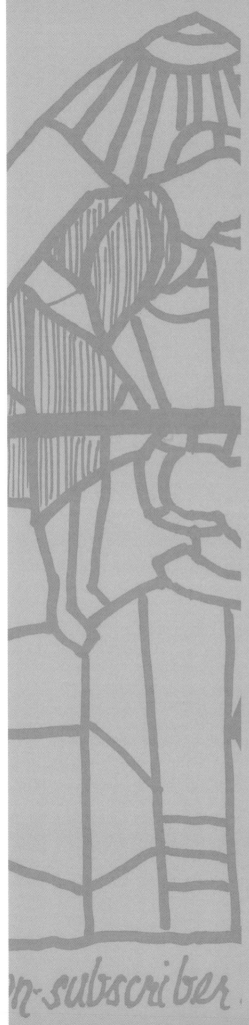

n-subscriber.

12: More Light, Less Heat
Domestic lighting layouts, installation and efficacy

Approved Document Part P: 'Design and Installation of Electrical Installations' came into force in April 2006 and directs designers to comply with BS 7671: 2001, the 16th edition of the IEE Wiring Regulations. Unsurprisingly, the 17th edition has been issued (BS 7671: 2008) and this Shortcut relates to current practice. It looks at domestic lighting, plugging into electrical regulations, standards and practices.

One week before Part P came into force, the Institute of Electrical Engineers merged with the Institute of Incorporated Engineers (IIE) to form the Institution of Engineering and Technology. However, whatever the name changes, it is the IEE Wiring Regulations manual that still remains as 'the bible' for any serious specifier of electrical work. This huge reference document provides detailed guidance for practically all aspects of electrical work, from initial safety considerations to the requirements for client maintenance.

Having an awareness of optimum lighting levels is important for a building's energy efficiency, comfort level and usability, but there have been a number of competing claims about the importance of well-lit buildings. Some have argued that, with the demise of the daylight protractor and the rise of the lighting consultant, building designers have lost their sense of control over the lighting that they deploy. Others say – tenuously, I contend – that the choice of lighting in non-domestic buildings is linked to the productivity and well-being of occupants. But uncontroversially, even in standard domestic installations, well-chosen and well-maintained lighting saves money. The government's Energy Efficiency Best Practice Performance Programme asserts that 'about 20 per cent of the electricity used in the UK is consumed by lighting' and a significant amount of that comes from buildings. Reducing that loading will ultimately reduce electricity bills. This Shortcut provides a basic understanding of the problems to be addressed when considering a lighting layout. First, let's consider the terminology.

Luminance is the measure of brightness – the strength of light in a given direction – and is measured in candelas/m².

LIGHTING LEVELS

Illuminance is the term given to the amount of light illuminating a given surface. It is also known as the 'task lighting level' and is measured in lumens/m², or lux: a lumen being the amount of luminous flux – or light – emitted from the source. Therefore, illuminance measures how much light reaches the relevant surface and should not be confused with 'luminance'. Luminance is the measure of brightness – the strength of light in a given direction – and is measured in candelas/m². The strength of the light tends to decline as the lamp gets older – and dirtier – which will inevitably detrimentally affect the illuminance, although new lamp technology (such as T5 tubes) and sensible cleaning regimes have made the fall-off in light output less of a problem than it used to be.

In general, illuminance refers to the lighting level at a desk or table height, except in corridors and stairs where it should be measured at the floor level. The Chartered Institution of Building Services Engineers' 'Code for Lighting' states that entrance lobbies to public buildings should be at around 200 lux, while the reception desk should be 500 lux. Banks, general offices and general purpose halls should be 300 lux while

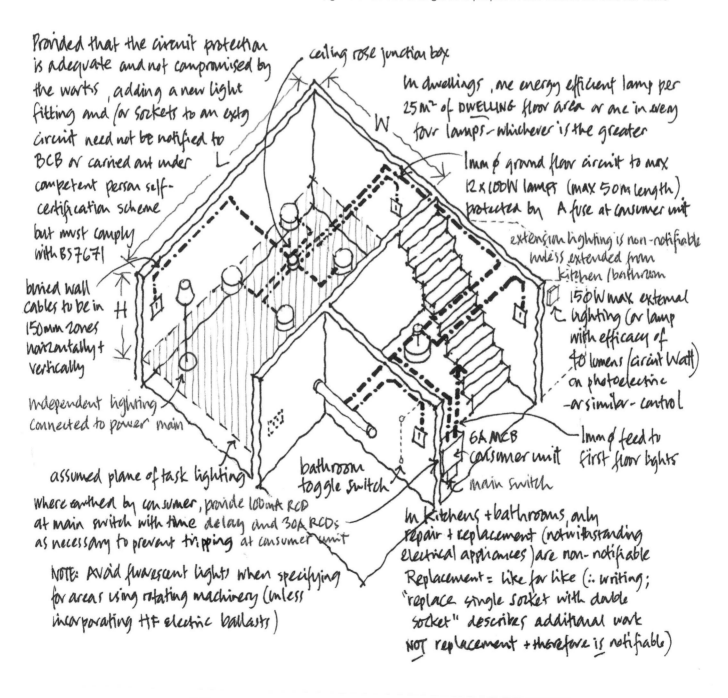

Provided that the circuit protection is adequate and not compromised by the works, adding a new light fitting and /or sockets to an extg circuit need not be notified to BCB or carried out under competent person self-certification scheme but must comply with BS7671

buried wall cables to be in 150mm zones horizontally + vertically

independent lighting connected to power main

assumed plane of task lighting

where earthed by consumer, provide 100mA RCD at main switch with time delay and 30A RCDs as necessary to prevent tripping at consumer unit

NOTE: Avoid fluorescent lights when specifying for areas using rotating machinery (unless incorporating HF electric ballasts)

ceiling rose junction box

In dwellings, one energy efficient lamp per 25m² of DWELLING floor area or one in every four lamps – whichever is the greater

1mm ø ground floor circuit to max 12 x 100W lamps (max 50m length) protected by A fuse at consumer unit

extension lighting is non-notifiable unless extended from kitchen / bathroom

150W max external lighting (or lamp with efficacy of 40 lumens /circuit Watt) on photoelectric – or similar – control

6A MCB consumer unit

1mm ø feed to first floor lights

bathroom toggle switch

main switch

In kitchens + bathrooms, only repair + replacement (notwithstanding electrical appliances) are non-notifiable Replacement = like for like (∴ writing; "replace single socket with double socket" describes additional work NOT replacement + therefore is notifiable)

workstations and public counters should be 500 lux. While shops range from 300 to 500 lux, supermarkets range from 500 to 1000 lux; corridors are set at 100 lux while electrical manufacturing facilities can be as high as 1500 lux. These are minimum levels sufficient to reduce risks to the users of those spaces, but much higher levels should be avoided to prevent the discomfort of glare and general overheating, the latter having an impact on the Dwelling Emission Rate (DER) in Building Regulations Approved Document Part L (see: Shortcuts: Book 2).

UNITS AND LOCATIONS

Lamps (commonly called 'light bulbs' or 'tubes') come in three main categories: tungsten filament, fluorescent and high intensity discharge lamps. The tungsten filament is the most common lamp used in domestic situations, but is also the least efficient in terms of the amount of light energy emitted per unit of energy inputted. Efficiency (also called efficacy) is measured in lumens per watt (lm/W) and a standard 100 W incandescent filament lamp provides around 1400 lumens, giving an efficacy of 14 lm/W. By comparison, a 58 W fluorescent tube provides around 5200 lumens (or 90 lm/W) and a 400 W high-pressure sodium lamp provides around 48,000 lumens giving an efficacy rating of 120 lm/W.

Compact fluorescent lamps are regularly used to replace the old tungsten filament lamps to comply with the Approved Document Part L requirement for more energy-efficient lamps. Older fluorescent tubes are not recommended for use in areas where rotating machinery is being used owing to the stroboscopic effect that the otherwise undiscernable pulsating fluorescent tube has on the machinery (see HSE document 'Lighting at Work'). Energy-efficient tubes with high-frequency electronic control gear – which are much more readily available than they were only a few years ago – should always be used in these circumstances.

Fluorescent induction lamps, which are still relatively unknown to the public, have been available since 1990 and often have a useful life of up to 60,000 hours or nearly 7 years of constant illumination. Because of their lifespan, they tend to be located in areas of limited accessibility. High intensity discharge (HID) lamps are very bright, although they take a little time to reach full intensity and have a tendency to get very hot and so, if located high up in naturally ventilated spaces, their impact on air flow patterns should be factored in.

To assess the number of lamps that will provide the task lighting level in a given situation, you need to calculate the Room Index (RI), which gives a clue to the Utilisation Factor (UF), which, in turn, is necessary to solve the equation to find the number of light fittings. The UF is the ratio between the lumens emitted and the lumens received at a location; the RI defines the relationship between the dimensions of the room. Calculating these numbers is relatively straightforward and is described in most lighting catalogues. The resulting data, when fed into most manufacturers' lamp/luminaire data, provides a basic rule-of-thumb guide to lighting compliance.

INSTALLATION RULES

The most common form of wiring layout for domestic socket outlets is called a ring circuit (or 'ring main'), in which a 2.5 mm² power cable, protected by 30 amp fuse (or miniature circuit breaker [MCB]) runs from the consumer unit connecting one socket to the next in a large circle (or ring) all the way back to the consumer unit, so that the start and finish points are at the same location.

$$\text{ROOM INDEX} = \frac{L \times W}{H(L+W)}$$

Using "room index" read off Utilisation Factor (UF) from manufacturers data specifying reflectances. Common reflectances are:
Ceiling (C): 70; Walls (W): 50; Floor (F): 20

$$\text{NUMBER OF LIGHT FITTINGS} = \frac{E \times A}{L \times UF \times M}$$

where, E = required illuminance (lux)
A = relevant area of task plane (say L x W)
L = lamp output (lumens), see manufacturer data
UF = Utilisation factor (see calculation above)
M = maintenance factor, see manufacturer data

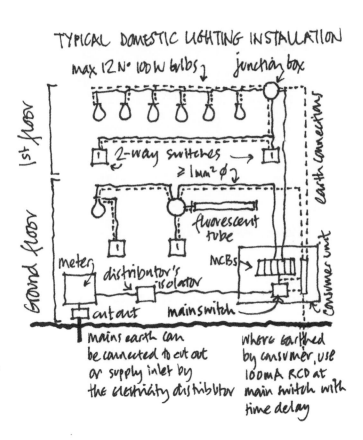

TYPICAL DOMESTIC LIGHTING INSTALLATION
max 12 N° 100W bulbs
junction box
1st floor
Ground floor
2-way switches ≥ 1mm² ∅
fluorescent tube
earth connections
consumer unit
MCBs
meter
distributor's isolator
cut out
main switch
mains earth can be connected to cut out or supply inlet by the electricity distributor
where earthed by consumer, use 100mA RCD at main switch with time delay

"

Residual circuit devices (RCDs) – sometimes called residual current circuit breakers (RCCBs) or residual current breakers (RCBs) – are all essentially trip switches that protect a circuit from power surges.

The 1 mm² cables connecting light fittings, however, are laid in a 'radial circuit' that, unlike a ring circuit, radiates from the consumer unit and terminates at the last light fitting. Domestic lighting mains are generally protected by a 6 amp fuse (or MCB). There are two types of radial system: Firstly, the 'loop-in method' has a single cable that runs from luminaire to luminaire, terminating at the last one with a cable run from each luminaire to its light switch(es). Secondly, the 'junction-box method', which tends to be found in older properties, incorporates a junction box at nodal points which have separate cables running to the luminaire and also to its switch(es). Both the loop-in and junction-box methods, or a combination of the two, are also acceptable in non-domestic situations (although the loop-in method has the advantage that all connections are readily accessible at the luminaires or ceiling rose, thereby making fault-finding easier). There should be no more than 6 m between the nearest luminaire and its control switch (although this may be extended to a maximum of two times the mounting height of the luminaire, if necessary). Bear in mind though, the requirements of the Disability Discrimination Act and specify the type and position of light switches, sockets, etc. so as to be easily accessible.

A single circuit of 1mm² PVC insulated and sheathed 2-core cable can generally be used to feed up to twelve 100 W light fittings in one run in typical domestic applications, although 1.5 mm² is also common. However, if lamps with a greater wattage or luminaries with multi-light fittings are included, then the circuit should be reassessed. If the cables are laid in situations that might cause them to overheat (located close to heating installations or grouped with other cables, for example) then the wiring specification should be upgraded. It is common for average-sized dwellings to have two separate lighting circuits serving upstairs and downstairs, but depending on the loading, additional circuits should be incorporated and clearly identified at the consumer unit.

Old-style cartridge fuses have generally been superseded with MCBs which 'trip' (cut out) when the current exceeds the safety standard of the circuit. However, where there is any possibility of metalwork becoming live due to a fault, the circuit must be protected with an earthed equipotential bonding and automatic disconnection of supply (EEBAD) so that the danger is cut off and residual current dissipated as quickly as possible, with the equipment connected to the earth and supplementary bonded.

In most practical applications, including domestic bathrooms, lighting circuits constitute a low risk and there is seldom a requirement to provide domestic lighting systems with residual circuit devices (RCDs) – sometimes called residual current circuit breakers, (RCCBs) or residual current breakers (RCBs) – which are all essentially trip switches that protect a circuit from power surges.

However, if the lighting is contained within 'special locations' (for example, swimming pools, saunas, non-traditional bathrooms, etc.) which are designated as 'high risk' areas, RCDs need to be considered for inclusion in lighting circuits. The high risk denotes the presence of water and the potential for earth leakage though moist air thus posing a clear risk to health and safety.

References

BS 8206-2 (1992) 'Lighting for Buildings. Code of Practice for daylighting'. BSI

Building Research Establishment (1996) Information Paper IP6/96: 'People and lighting controls', BRE.

BS 7671 (2001) 'Requirements for Electrical Installations: IEE Wiring Regulations', 16th edn, The Institution of Engineering and Technology (no longer current but cited in the Building Regulations).

BS 7671 (2008) 'Requirements for Electrical Installations: IEE Wiring Regulations', 17th edn, The Institution of Engineering and Technology.

Building Research Establishment (2001) BRE Report BR430 'Energy efficient lighting: Part L of the Building Regulations Explained', BRE.

Department for Communities and Local Government (2006) 'Approved Document Part P: Design and installation of electrical installations', NBS.

Health and Safety Executive (1997) Health and Safety Guide 38: 'Lighting at Work', HSE.

Mardaljevic, J. (2006) Building Services Journal 09/06 'Time to see the light', pp: 59–62.

RECOMMENDED READINGS

Chartered Institution of Building Services Engineers (2002) 'Code for Lighting', Parts 0, 1, 2, 3 & 4, CIBSE.

Energy Savings Trust (2006) 'GIL 20, Low Energy Lighting', EST.

Saules, T.D. (2001) 'The Illustrated Guide to Electrical Building Services', BSRIA.

13: Rainscreen Cladding
Letting air in, to keep rain out

Rainscreen cladding has been around, in various guises, for centuries, but its modern, scientifically validated incarnation was developed in Scandinavia during the 1940s. During the 1950s, the UK's Building Research Station declared the advantages of drained and ventilated air spaces behind impervious outer envelopes. However, it took a further 25 years for it to become commonplace.

The Centre for Window and Cladding Technology (CWCT) 'Standard for systemised building envelopes' defines curtain walling as 'a form of predominantly vertical building envelope which supports no load other than its own weight and the environmental forces which act upon it'. Similarly, rainscreen cladding is a non-loadbearing external cladding assembly defined as 'a wall comprising an outer skin of panels and an airtight insulated backing wall separated by a ventilated cavity. Some water may penetrate into the cavity but the rainscreen is intended to provide protection from direct rain.' So, the key distinction is that curtain walling is usually the whole envelope, while rainscreen cladding is the outer protective layer of the envelope.

As far back as the Second World War, UK researchers were exploring the potential of a protective impervious cladding applied to the outer face of external walls with a drained and ventilated air gap behind it. In 1952, the Alcoa building in Pittsburgh – a 30-storey building clad in open-jointed aluminium baffle panels – became one of the earliest and best-known examples of rainscreen cladding, even though the phrase hadn't yet been invented at the time of its construction (the term 'rainscreen *principle*' was coined in the

In 1962, the Norwegian Research Institute pioneered rainscreen cladding... In 2008, the Department for Communities and Local Government discovered the 'innovative' nature of rainscreen cladding.

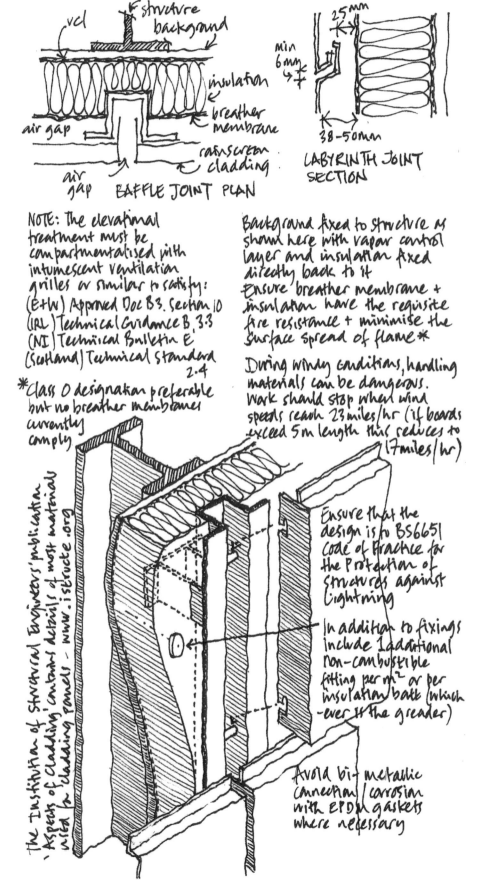

vcl

structure background

insulation

breather membrane

air gap

rainscreen cladding

air gap

BAFFLE JOINT PLAN

25mm

Min 6mm

38-50mm

LABYRINTH JOINT SECTION

NOTE: The elevational treatment must be compartmentalised with intumescent ventilation grilles or similar to satisfy:
(E+W) Approved Doc B3. Section 10
(IRL) Technical Guidance B 3·3
(NI) Technical Bulletin E'
(Scotland) Technical Standard 2·4

*Class O designation preferable but no breather membranes currently comply

Background fixed to structure as shown here with vapour control layer and insulation fixed directly back to it
Ensure breather membrane + insulation have the requisite fire resistance + minimise the surface spread of flame*

During windy conditions, handling materials can be dangerous. Work should stop when wind speeds reach 23 miles/hr (if boards exceed 5m length this reduces to 17 miles/hr)

The Institution of Structural Engineers' publication 'Aspects of Cladding' contains details of most materials used for cladding panels – www.istructe.org

Ensure that the design is to BS6651 Code of Practice for the Protection of Structures against Lightning

In addition to fixings include 1 additional non-combustible fitting per m² or per insulation batt (whichever is the greater)

Avoid bi-metallic connection/corrosion with EPDM gaskets where necessary

1970s). It was in 1962 that the Norwegian Research Institute published a booklet which championed systems that ensure pressure equalisation of the air gap (that means that the air pressure in the gap separating the rainscreen cladding from the structure is the same as that of the outside conditions). This simple pressure equalisation principle was shown to combat wind-driven rain infiltration. In 2008, the Department for Communities and Local Government included rainscreen cladding as an 'innovative' technique.[1]

Since then, the two systems – 'drained and ventilated' and 'pressure equalised' – comprising a notionally impervious and a permeable outer layer respectively, have been widely specified. The degree of acceptable water penetration into the air gap defines the difference between the two systems. In a drained and ventilated system it is assumed that, even though it is notionally impervious, rain, snow or hail can still be wind-driven into the air gap, and therefore careful detailing is required in order to prevent water crossing the gap and penetrating the structure beyond.

In a pressure-equalised system, the relationship between the area of the open joint which gives access to the gap, the volume of the gap and the air permeability of the air barrier is designed such that wind pressure acting on the face of the rainscreen is balanced by the pressure created at the joint. The air gap acts as a pressure cushion to prevent water from reaching the insulation and backing wall and, as such, the air gap pressure is an essential component preventing excessive water passing through an open jointed rainscreen (Note: 'Open joints' include baffled and labyrinth joints). As it happens, drained and ventilated rainscreens also achieve a certain degree of pressure equalisation.

To add to the confusion, the term 'open-jointed rainscreen' has emerged to describe both systems, and the understanding of the two technologies has become ill-defined. Currently, there is no specific British Standard for ventilated rainscreen walls. BS 8200 'Code of practice for the design of non-loadbearing external vertical structures' includes general details about the principles of drained and ventilated and pressure-equalised systems. Published in 1985, with no revisions, the standard is categorised by BSI as current but obsolescent, as it no longer

A typical examination of dewpoint calculations suggests that condensation will form *within* a material – for example, within the body of the insulation – but in reality, especially if the insulation is nominally vapour impermeable, condensation actually tends to form at the interface of the materials, along the surface plane. That is, it forms in the interstice of abutting materials. A well-designed vapour control layer (VCL) should prevent the passage of moisture into the building fabric.

PROBLEM AREAS

The fact that gaseous water vapour condenses into beads of moisture may only be a problem if the diffusion of that moisture, or its evaporation, is insufficient to prevent its build-up; that is, if it forms more quickly than it re-evaporates. The core problem of interstitial condensation is this gradual increase in 'trapped' moisture, which can contribute to:

- timber decay
- metal corrosion
- dimensional changes
- salts migration
- electrical failure
- staining
- the reduction in the thermal performance of the insulation.

Care must be taken when using UK-centred specifications to design buildings for warmer climates (or when designing buildings with extensive air conditioning systems in the UK), because reverse vapour flows have to be taken into consideration. In these cases, positioning a VCL on the internal side of the insulation – i.e. as normal for UK detailing, but in this instance potentially on the cooler side of the insulation – means that water may condense at the interstice between the insulation and the VCL, and so the VCL must be repositioned to prevent it.

Approved Document C 'Resistance to moisture and weather' (AD C) provides no direct information on how to avoid interstitial condensation, suggesting only that 'a ground floor or floor exposed from below, i.e. above an open parking space or passageway' an 'external wall' or 'roof' be designed to BRE BR262, BS 5250 and BS EN ISO 13788. Special attention should be given to those buildings which

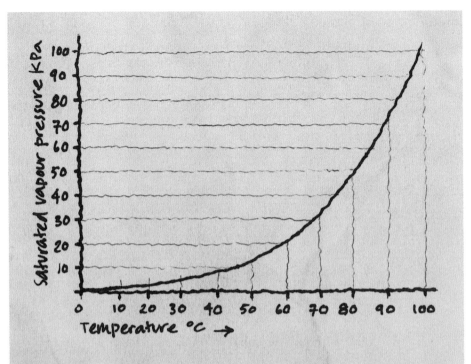

Diagram 3:
When the saturated vapour pressure of a liquid becomes equal to the external pressure on the liquid, it will boil. This graph shows how the svp of water varies between 0°C and 100°C.

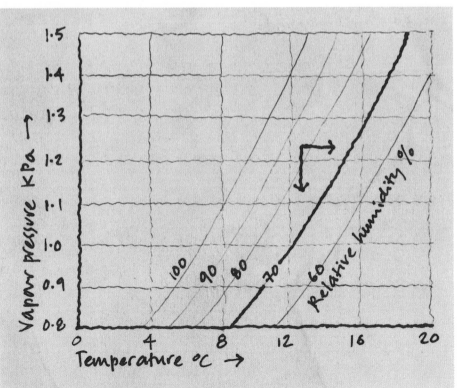

Relative humidity, vapour pressure and temperature
Technical information taken from BS 5250:2002

Approved Document C 'Site preparation and resistance to contaminants and moisture' provides no direct information about how to avoid interstitial condensation.

have, by dint of their operation, high moisture content. AD C also states that 'the requirement will be met by the ventilation of cold roofs'. Published three years later, STH3 suggests that 'cold, level-deck roofs should be avoided because interstitial condensation is likely'. There is no mention of interstitial condensation in any of the other Building Regulations Approved Documents, but the essence of reasonable and effective interstitial condensation prevention is the inclusion of a vapour control layer on the warm side of the construction to prevent moisture vapour penetration to a position where it will reach dewpoint.

DEFINITIONS

Dewpoint – Temperature at which air becomes saturated with water vapour.

Vapour pressure – Part of atmospheric pressure due to water vapour present in the air, expressed in kPa (1 kPa = 10 mbar = 1000 N/m²).

Saturation vapour pressure – Water vapour pressure in air at dewpoint temperature.

Relative humidity – The ratio of the vapour pressure in air at a given temperature to the saturation vapour pressure at the same temperature, commonly expressed as a percentage.

Vapour resistance – This is the measure of the resistance to water vapour diffusion of a material or combination of materials of specific thickness, expressed in MN.s/g (mega-Newton seconds per gram). Note: For membranes, the performance is stated as vapour resistance. For other materials, it is obtained by multiplying thickness by vapour resistivity (see below). As an example, Tyvek is a vapour permeable material of 0.23 MN.s/g (manufactured by DuPont – pronounced 'dewpoint').

Vapour control layer – Material that substantially reduces the water vapour transfer through any building component in which it is incorporated by limiting both vapour diffusion and air movement. The Scottish Building Standards Agency states that 'a vapour control layer is a material with a vapour resistance greater than 200 MN.s/g'. BS EN 1931, 'Flexible sheets for waterproofing – Bitumen, plastic and rubber sheets for roof waterproofing – determination of water vapour transmission properties' rates polyethylene, for example, at 250 MN.s/g.

Vapour resistivity – This is the measure of resistance of a unit thickness of material to water vapour diffusion expressed as MN.s/g. Note: For membranes, the performance is stated as vapour resistance (see above).

Reverse condensation – Interstitial condensation formed by water vapour travelling from outside to inside, i.e. the reverse to 'normal' flows.

Breather membrane – A relatively permeable membrane having a vapour resistance of less than or equal to 0.6 MN.s/g – see definition of 'vapour resistance.

References

BRE (2004) *'Safety and Health in Buildings: Advanced calculations of moisture movement in structures'*, BRE Scotland.

BS 6229 (2003) *'Flat roofs with continuously supported coverings. Code of practice'*, BSI.

Flat Roofing Alliance (1999) Information Sheet No. 6: *'Condensation & Vapour Control Layers'*, FRA.

Office of the Deputy Prime Minister (2004) *'The Building Regulations 2000: Approved Document C: Site preparation and resistance to contaminants and moisture'*, NBS.

RECOMMENDED READINGS
BRE (2002) BRE 262 *'Thermal insulation: avoiding risks'*, 3rd edn, BRE.

BRE (2005) Information Paper *'Modelling and controlling interstitial condensation in buildings'*, BRE.

BS 5250 (2002) *'Code of Practice for control of condensation in buildings'* (AMD 16119), BSI.

BS EN ISO 13788 (2002) *'Hygrothermal performance of building components and building elements. Internal surface temperature to avoid critical surface humidity and interstitial condensation. Calculation methods'* (AMD Corrigendum 13792), BSI.

Scottish Executive (2007) *'Scottish Technical Handbook: Domestic 3: Environment'*, SE.

Scottish Executive (2007) *'Scottish Technical Handbook: Non-Domestic 3: Environment'*, SE.

18: Wooden Performance
Moisture content of timber

Timber is one of the most ubiquitous products used in construction, but in order to perform adequately it needs to be carefully prepared, stored and maintained. To ensure satisfactory preservative treatment, accurate machining and efficient fabrication, and to avoid problems due to dimensional change and distortion in use, its moisture content must be controlled.

BS 6100-1 (also numbered BS ISO 6707-1), 'Building and civil engineering – Vocabulary', describes joinery (known as 'cabinetry' in America) as the assembly of worked components of timber and wood-based panels other than structural timber or cladding, together with associated mouldings used as finishing, such as architraves, skirting boards, weatherboards, etc. Carpentry, on the other hand, is defined as 'structural woodwork'. Woodwork is quite simple: it means working with wood.

As an aside, BS 6100-1 also contains several words used in the US construction industry and the equivalent 'preferred international terms' for the same things. So, for instance, where Americans say 'mound', we are told that Europeans say 'dumpling'; where we say 'pugging', Americans say 'deafening fill'; where Americans say 'stabilization', allegedly Europeans say 'structural rehabilitation'. As Oscar Wilde might have said, two nations separated by a common language.

A piece of timber weighing 500 g, and containing 250 g of water will have a moisture content of 100 per cent.

In Shortcuts: Book 2 we explore the reality of harmonised regulations and standards, but one of the latest impositions of harmony on UK construction practices is the new edition of EN 942, 'Timber in Joinery', which came into force in September 2007 alongside three other documents (see Recommended Readings). These superseded conflicting national standards, which were effectively withdrawn in October 2007, although decisions on issues such as the moisture content of timber will continue to be influenced by relevant product standards or national requirements until such time as blanket European coverage is imposed.

The moisture content of wood is normally expressed as a percentage, and is calculated as: the difference between the weight of a sample of 'wet' (green) wood and the weight of the same sample after oven drying (to remove all moisture), divided by the oven-dry weight, all multiplied by 100. Thus, a piece of timber weighing 500 g, and containing 250 g of water will have a moisture content of 100 per cent or [(500–250) / 250] × 100.

The recommendations given in EN 942 for moisture content at the completion time of the product manufacture are useful where specific national product standards, etc. do not exist. It states that, *at the time of manufacture*, external joinery should have a moisture content *in use* of 12–19 per cent and that the moisture content of 'internal joinery' should be:

■ 12–16 per cent in unheated buildings

■ 9–13 per cent in heated buildings between 12–21°C

■ 6–10 per cent in heated buildings with temperatures in excess of 21°C.

These values vary slightly from UK national product standards, which typically recommend that *at the time of installation* the moisture content should be:

■ 18 per cent in covered, generally unheated spaces

■ 15 per cent in covered, generally heated spaces

■ 12 per cent in internal conditions, in continuously heated buildings

■ 20 per cent or more for external timber.

According to BS 5268-2 'Structural use of timber. Code of practice for permissible stress design, materials and workmanship', if a moisture content lower than 18 per cent is required at the time of erection, extra drying costs should be anticipated. Timber on site must be adequately protected, both before and after installation, to prevent the moisture content from rising. This is particularly important with material dried to below 20 per cent moisture content since the full design load should not be applied if the moisture content rises above 20 per cent.

The strength of timber is also affected by its moisture content – increasing as the moisture content reduces and vice versa. For example, the bending and the compression stresses for 'wet' timber, i.e. wood with a moisture content exceeding 20 per cent, are 80 per cent and 60 per cent respectively of those for 'dry' timber (wood with a moisture content less than 20 per cent). Not only that, but the thickness and width of a piece of timber is predicted to increase by 0.25 per cent for every 1 per cent increase in moisture content above 20 per cent, and to shrink by the same amount for every 1 per cent reduction in moisture content below 20 per cent. Therefore, it is essential that structural timber (not joinery timber) be strength graded at a moisture content appropriate for the exposure conditions of the timber in use. Three service classes are defined in BS 5268-2:

Service class 1 This is characterised by a moisture content in the materials corresponding to a temperature of 20°C, and the relative humidity of the surrounding air only exceeding 65 per cent for a few weeks in the year. In such conditions most timber will attain an average moisture content not exceeding 12 per cent.

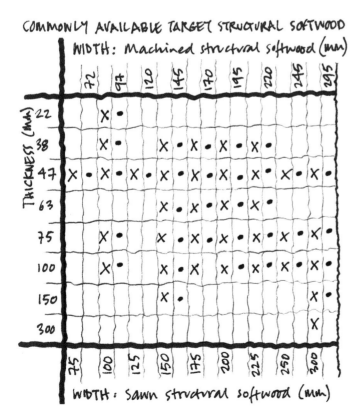

COMMONLY AVAILABLE TARGET STRUCTURAL SOFTWOOD

WIDTH: Machined structural softwood (mm)

WIDTH: Sawn structural softwood (mm)

Service class 2 Having a moisture content in the materials corresponding to a temperature of 20°C, and when the relative humidity of the surrounding air only exceeds 85 per cent for a few weeks in the year. In this instance, most timber will attain an average moisture content not exceeding 20 per cent.

Service class 3 This is for materials with a higher moisture contents than Service class 2.

Timber with a moisture content at or below 20 per cent at the time of grading to BS 4978 or BS EN 14081-1 must be stamped 'DRY' or 'KD' (kiln dried). This is the average moisture content (the maximum moisture reading must not exceed 24 per cent). Softwood with an average moisture content above 20 per cent at the time of grading must be stamped 'WET' and must be used only in high-moisture environments unless the design takes it into account. Some North American timber may be marked 'GRN' (green or unseasoned) which is equivalent to our 'WET'.

	Type	Workability	Durability	Suitability			Movement values
				External joinery		Internal joinery	movement at change in humidity from 90–60% at constant temp
				Cills, threshold, door frames	Windows, fascias, etc.		
Softwoods	Douglas Fir	A	S	NP	NP	NP	under 3%
	Parana Pine	E	M	X	X	NP	3–4.5%
	Southern Pine	E	M	P	P	NP	n/a
	Redwood: Scots Pine	A	M	P/P	P	NP	3–4.5%
	Western Red Cedar	A	S	NP/P	NP	NP	under 3%
	European Whitewood	E	M	P/P	P	NP	3–4.5%
Hardwoods	American Ash	A	S	X	X	NP	3–4.5%
	European Ash	A	M	X	X	NP	3–4.5%
	European Beech	E/A	L	X	X	NP	over 4.5%
	Central/S American Cedar	A	S	NP/P	NP	NP	under 3%
	American Cherry	A	M	X	X	NP	3–4.5%
	Sweet Chesnut	A	S	NP	NP	NP	over 4.5%
	Elm	A	M	X	X	NP	3–4.5%
	Iroko	A/D	S	NP	NP	NP	n/a
	Lauan	E/A	S	P	P	NP	under 3%
	African Mahogany	A	S	NP	NP	NP	under 3%
	American Mahogany	A	S	NP	NP	NP	under 3%
	Makore	D	S	NP	NP	NP	under 3%
	Soft Maple	A	M	X	X	NP	3–4.5%
	American Red Oak	A	M	X	X	NP	3–4.5%
	European Oak	A	M	NP	NP	NP	3–4.5%
	Japanese Oak	A	L	X	X	NP	3–4.5%
	Sapele	A	M	NP	NP	NP	3–4.5%
	Sycamore	A	M	X	X	NP	n/a
	Teak	D	S	NP	NP	NP	under 3%
	American Walnut	A	S/M	NP	NP	NP	3–4.5%

KEY: A=easy; **B**=average; **C**=difficult; **S**=small; **M**=medium; **L**=large; **NP**=suitable without preservative; **NP/P**=suitable without preservative if covered*; **P**=suitable if preservative treated to BS 5589; **P/P**=suitable if preservative treated to BS 5589 and covered*; **X**=unsuitable. *cover to prevent mechanical damage at cills, etc.

EUROPEAN AND US TERMS

'Dumpling' (European) or 'mound' (US) when referring to a large mass of ground intended to be excavated, but temporarily left as a support during construction work.

'Pugging' or 'deafening fill' is sand or other similar material used above ceilings, between joists, to assist in sound insulation.

'Structural rehabilitation' or 'stabilization' describes the application of measures designed to re-establish the structural stability, functionality or both of a building and its enclosure, while essentially maintaining the existing form.

A NOTE ON SOME PRESERVATIVES

The approval for copper chrome arsenic (CCA) preservative treatment was withdrawn on 1 September 2001, but CCA-treated timbers may still be imported. The Controls on Dangerous Substances and Preparations Regulations 2006 state that these CCA timbers may only be used if they contain 'Type C' solutions (i.e. a diluted solution, typically containing no more than 3 per cent preservative ingredients, with 97 per cent or more water) and are used in industrial installations using vacuum or pressure to impregnate wood. Wood so treated may not be placed on the market before fixation of the preservative is completed. To safeguard users and the public, all CCA-treated wood placed on the market must be labelled 'For professional and industrial installation and use only – contains arsenic'. Wood and waste wood treated with this preservative should be regarded as being hazardous and must be disposed of by an authorised undertaking.

Creosote has been banned in the UK since 2004, and it is now illegal even to keep a tin of the stuff. It can only be obtained in special 20+ litre containers for professional pressure-applied uses. There is a market for second-hand creosoted timbers, but none may be used inside buildings, in playgrounds, parks, gardens and outdoor leisure facilities where there is a risk of frequent skin contact or in other situations where the creosote might come into contact with skin, clothes or products destined for human and/or animal consumption.

There is an increasingly specialist laminate market using timber which, during the manufacturing process, has a very high moisture content. In conventional operations, timber for laminates is first dried before defects are cut out and the resulting fault-free lengths are finger-jointed together to build up an engineered laminate. Much of this timber is thus discarded *before* manufacture but *after* drying, meaning that a lot of energy has been expended in drying waste timber. The 'Greenweld' process initially developed and patented by the New Zealand Forest Research Institute (NZFRI) bonds the laminate prior to drying. It utilises low-grade timber with a moisture content of between 18–180 per cent. In this way, kilns could be loaded efficiently to maximum capacity with long, fully saleable lengths of quality product. In addition, the process increased productivity by making use of otherwise-useless short lengths of timber.

Unfortunately, the original companies behind the Greenweld™ process have sold up or gone into liquidation, but research is continuing in Europe to explore these opportunities.[1] Not least, research is being carried out at Wood Knowledge Wales in Bangor.[2] As yet, in the UK, there are no standards governing this method of manufacture, and BRE has suggested that it should be used for select, non-structural work only.[3] 'Wood Material Science and Engineering' magazine advises that 'studies of green finger-jointing made with different adhesives have shown improved properties compared with dry-glued joints, but have not provided explanations for the improvement'.[4] Experiments with this technology should be carried out with caution.

[1] Morlier, P. & Pommier, R. (2006) *'Finger jointing on green maritime pine timber: Improving the process and final performances'*, Laboratoire de Rhéologie du Bois de Bordeaux, (UMR 5103). See: http://www.ewpa.com/ Contact: pommier@lrbb3.pierroton.inra.fr

[2] Wood Knowledge Wales, Biocomposites Centre, University of Wales. www.woodknowledgewales.co.uk

[3] Cooper, G., Cornwell, M. & Thorpe, W. (2005) *'Green gluing of timber: a feasibility study'*, BRE

[4] Elbez, G. & Pommier, R. (2006) *'Finger-jointing green softwood: Evaluation of the interaction between polyurethane adhesive and wood'*, Wood Material Science and Engineering, 1: 127A137

References

BS 4978 (2007) *'Visual strength grading of softwood – Specification'*, BSI.

BS 5268-2 (2002) *'Structural use of timber – Part 2: Code of practice for permissible stress design, materials and workmanship'* (incorporating Amendment No. 1), BSI.

BS 6100-1 (also labelled BS ISO 6707-1) (2004) *'Building and civil engineering – Vocabulary – General Terms'*, BSI.

BS EN 350 *'Durability of wood and wood-based products – Natural durability of solid wood – Part 2: Guide to natural durability and treatability of selected wood species of importance in Europe'*, BSI.

BS EN 519 (1995) *'Structural timber. Grading. Requirements for machine strength graded timber and grading machines'*, BSI.

Cooper, G., Cornwell, M. & Thorpe, W. (2005) *'Green gluing of timber: a feasibility study'*, BRE.

TRADA Technology (2006) *'Moisture in timber'*, Section 4 Sheet 14, TRADA.

RECOMMENDED READINGS

BS 5268-2 (2002) *'Structural use of timber – Part 2: Code of practice for permissible stress design, materials and workmanship' (incorporating Amendment No. 1)*, BSI.

BS EN 942 (2007) *'Timber in joinery – General requirements'*, BSI.

BS EN 13307-1 (2007) *'Timber blanks and semi-finished profiles for non-structural use – Part 1: Requirements'*, BSI.

BS EN 14220 (2007) *'Timber and wood-based materials in external windows, external door leaves and internal doorframes – Requirements and specifications'*, BSI.

BS EN 14221 (2007) *'Timber and wood-based materials in internal windows, internal door leaves and internal doorframes – Requirements and specifications'*, BSI.

19: You've Been Framed Specification data for glass and glazing

BS 952-1: 1995 'Glass for glazing' lists over 20 main varieties of glass, with many more subsets representing composites, coated glass or other manufacturing processes that give the glass a range of properties suited to particular situations. This Shortcut looks at some of the types of glass that are available.

There are some British Standards that are fascinating, well researched, informative and crucial for an understanding of a subject, product or practice ... and then there's BS 6262-1: 2005. Admittedly, 'Glazing for Buildings – General methodology for the selection of glazing' is the first part of a seven-part document but, containing nuggets such as 'in the case of internal doors, the wind loading is negligible' and 'security glazing is used in situations where a high degree of protection ... is required', surely few people would weep if it disappeared. Fortunately, the remaining six parts do provide a more valuable resource, with detailed information on glass- and glazing-related topics, from energy considerations to frame design. Here we present a snapshot of available glazing types suited to different functions, locations and manufacturing processes.

ORDINARY GLASS
Henry Bessemer, more famous perhaps for his iron-to-steel converter, patented float glass in 1848, although it was not until the turn of the 20th century that Pilkington made it commercially viable to produce it in any significant quantities.

Heat-soaking is a quality control process that 'soaks' the glass at a temperature of around 300°C for several hours to speed up the rate of expansion of any nickel sulfide inclusions that might be present.

Glass does not weaken or degrade over time. Windows that flex under loads, for example, do not suffer repetitive stress fatigue.

It is manufactured, as the name suggests, by pouring 1000°C glass onto a bath of molten tin so that it floats. The glass spreads out and reaches equilibrium at an even thickness of around 7 mm. Thinner sheets (down to 0.4 mm) are produced by removing it slowly, thus allowing it to spread further; thicker sheets (up to 25 mm) are produced by removing it from the bath more quickly and thereby containing the spread. The industrialised efficiency heralded by the early manufacture of float glass, which became fine tuned after the Second World War, has now resulted in around 260 float glass plants worldwide, currently producing around 800,000 tonnes of glass per week between them. Other methods of glass production still exist, but float glass is generally accepted to be the most cost-effective.

For example, time-consuming and costly plate-glass production involves pouring a pool of molten glass onto two metal plates and allowing it to cool. This is then lifted off and ground and polished in an attempt to make its two surfaces even.

The other main glass production process involves passing molten glass through rollers of various apertures to control its thickness, thus producing 'sheet glass'. This glass tends to have very uneven surfaces and its production has effectively been superseded by float glass.

Glass that has been subjected to controlled cooling, which helps to reduce the residual stresses within the material, is known as 'annealed glass' or 'untreated glass'. Because the thermal conductivity of glass is so low, anyone who has put a hot glass in cold water will know what uncontrolled cooling can do when the stresses within the material are not contained. BS 952-1: 1995 describes annealed glass as 'ordinary glass' and it forms the base material for most wired, coloured and patterned glass produced today.

Glass is remarkable for its ability to not weaken or degrade over time. Windows that flex under wind loads, for example, do not weaken and fail due to repetitive stress fatigue. Unlike steel where regular flexure can cause permanent deformation, the molecular structure of glass is such that it retains its shape after flexing, except under excessive strain when it will fail suddenly and dramatically. Ordinary annealed glass is not very strong and a range of treatments can be introduced during its manufacture in order to moderate the tendency for float glass to break into dangerous fragments.

Some of the key techniques for improving the strength of glass include toughening, heat strengthening and laminating. It was early in the mid 17th century that heat-treated glass was discovered to have significantly improved strength qualities, but it took another two hundred years to be able to explain it scientifically. Essentially, rapid cooling causes the exterior of the glass to solidify very quickly, while the core takes longer. The core thus contracts to a greater extent than the surface, setting up tensile stresses in the core and compressive strength at the surface. It is the latter that has to be overcome before the 'weakness' of the internal condition can be exposed. There are two main types of heat-treated glass commercially available – toughened and heat-strengthened.

TOUGHENED GLASS

Toughened glass (also known as tempered glass) has a surface compressive strength of up to 100 N/mm² suitable for safety glass specifications, and can be as high as 150 N/mm² enabling it to withstand mechanical forces and thermal shock, respectively four and six times that of annealed glass (although this does not mean that it has enhanced fire safety characteristics).

Unfortunately, toughened glass is no more elastic than annealed glass and its deflection characteristics are limited. Its

	FRAME	FIRE PERFORMANCE TO BS EN 13501		SAFETY CLASSIFICATION		BREAKAGE
		Duration	Behaviour	Thickness (mm)	Class	
ordinary wired annealed glass	wood	≤ 60	Glass fractures but the shards held in place by wire mesh thus maintaining smoke + flame barrier	6mm cast 7mm polished	None or Class B	Glass fractures but held in place by wire mesh
	steel	≤ 60				
safety wired annealed glass	wood	≤ 60	As above	6mm cast 7mm polished		As above but safety wired glass may improve classification
	steel	≤ 60			Class C	
3-ply laminate: ordinary wired glass (or safety wired)/PVB/float glass layers	wood	see manufacturers data	Float glass fractures. PVB interlayer melts + burns away allowing float glass shards to fall + exposing the wired glass to fire	8.5mm and over	Class A N Class B depending on thickness	
	steel	ditto		ditto		ditto

BS 6206 'safety glass' label is premised on the fact that, on breaking, the release of pent-up internal energy shatters the glass into small, blunt, relatively harmless particles.

Such are the internal forces, that pane edges must be polished to remove irregularities which would otherwise reduce the stress resistance. These glasses should not be considered for security glazing. They cannot be cut or drilled after tempering, so manifestation and other decorative treatments are usually applied rather than etched, although the latter is possible with care.

Spontaneous fractures due to nickel sulfide inclusions have been overstated in recent years and are a relatively rare occurrence. It refers to the slow expansion of imperfections of nickel sulfide within the core of the glass that form during the manufacturing process. This expansion is caused by the reaction of the nickel sulfide to heat – traditionally from sunlight after the panel has been installed – which increases the volume of the nickel sulfide crystals sufficiently to upset the balance of stresses within the glass, resulting in a local fracture which instantaneously propagates throughout the pane. Heat-soaking is a quality control process that 'soaks' the glass at a temperature of around 300ºC for several hours to speed up the rate of expansion of any nickel sulfide inclusions that might be present. This is an expensive safeguard which results in the failure of defective panels prior to delivery and also induces failure in panes that may be substandard in other ways, e.g. those with edge defects. Because a 'successful' test will result in the destruction of the glass, it is generally only used for high-risk glazing and only on a random percentage of the panels.

HEAT STRENGTHENED GLASS

Heat strengthened glass is simply a half-way house between annealed and toughened glass, meaning that it is cooled faster than ordinary glass, but more slowly than toughened glass. It produces panels with a surface compressive strength of 25–60 N/mm^2 and a resistance to mechanical forces and thermal shock respectively more than 1.5 times and 2 times that of ordinary glass. Unfortunately, when it breaks, it fractures into shards like annealed glass.

WIRED GLASS

Wired glass, commonly known as Georgian wired glass, is annealed glass that has welded wire mesh embedded in it during the rolling process (used to create the desired patterning). Outside the UK it is made by pressing the mesh into the molten glass. Cast (patterned) wired glass is 7 mm thick, and the polished wired glass is made by grinding and polishing cast glass to a thickness of 6 mm.

In general, standard off-the-shelf panes of wired glass *cannot* be used as safety glass because the shards, when broken and combined with the sharp mesh, present a serious injury hazard. However,

delivery of raw materials or cullet (waste)

NOTE: there's 266,000 tonnes of waste float glass in UK/annum

melting furnace to 1000+°C

molten tin

25mm glass made if removed quickly

cooling "lehr": the controlled cooling kiln that reduces the temperature from 1000 to 200°C and creates annealed glass sheet

float bath

glass settles to 7mm thickness on bed of molten tin

annealed (ordinary) glass panes with compressive strength 25 N/mm2

continuous ribbon of glass on conveyor

cutters + cross cutters

finished panes for offloading + delivery

toughened glass has compressive strength 100 - 150 N/mm²

with a thicker gauge mesh it can be made to comply and can also satisfy BS 6206 class C safety classification (the lowest grade of safety glazing). For glazing in doors and side panels in critical locations, as defined in Approved Document Part N, the width of each pane of Class C glazing must not exceed 900 mm. Wired glass can also provide up to 60 minutes fire resistance (integrity only).

LAMINATED GLASS

Laminated glass is made by pressure-bonding sheets of glass to an interlayer, most commonly a polyvinyl butyral (PVB) resin, such that the composite panel has the combined attributes of each individual layer. Laminated glass using panes of float glass can be worked after manufacture, as the strength of the bond between the layers is unaffected by cutting and chasing. Note that laminates made from two toughened glass panels (known as 3-ply) will afford no structural integrity if both panels are broken.

PVB comes in thicknesses in multiples of approximately 0.38 mm. 3-ply composites with a 0.38 or 0.76 mm interlayer are classified as safety glasses. A 3-ply composite with a 1.53 mm interlayer (or a 5-ply with two 1.14 mm interlayers) qualifies as manual attack-resistant glass to BS 5544. Bullet and blast-resistant glass are specified in BS EN 1063: 2000, the former weighing as much as 120 kg/m^2 for a 50 mm thick pane.

A variety of laminates is now readily available providing intumescent, coloured or solar control interlayers. Ultraviolet-resistant interlayers, for example, can absorb 99 per cent of normal range ultraviolet radiation while allowing the light through relatively unhindered. However, the coloured, intumescent and solar control interlayers, as well as the coated and tinted glass laminates, tend to have a detrimental effect on light transmittance through the composite panels. An alternative laminated glass manufacturing process uses resins instead of PVB film interlayers.

FIRE-RESISTANT GLASS

In the same way that sound resistance relates to a composite construction (e.g. a glazed screen) rather than a component (e.g. an individual glass pane) so it is with fire resistance: failure of one part of the element compromises the whole. The fire resistance of a glazed screen, say, is a mixture of integrity and insulation – the former ensures that the fire does not penetrate the construction; the latter reduces the transmission of radiant and conductive heat.

Some fire-resistant glass is classified in accordance with BS EN 13501-2 in terms of integrity (E), insulation (I) and, where applicable, radiation (W). Note that ratings given on fire test certificates apply only to that particular assembly tested. It is important to note that fire protection is related to the extant conditions, surrounds, materials and detailing and, as such, claims that a certain glass is fire resistant should be treated with caution until read in conjunction with the whole assembly.

References

BS 952-1 (1995) 'Glass for glazing – Part 1: Classification', BSI.

BS 952-2 (1980) 'Glass for glazing – Part 2: Terminology', BSI.

BS 5544 (1978) 'Specification for anti-bandit glazing (glazing resistant to manual attack)' (AMD 4762). BSI.

BS 6206 (1981) 'Specification for impact performance requirement for flat safety glass and safety plastics for use in buildings' (AMD 4580) (AMD 5189) (AMD 7589) (AMD 8156) (AMD 8693), BSI.

BS 6262-1 (2005) 'Glazing for buildings – Part 1: General methodology for the selection of glazing', BSI.

BS EN 572-1 (2004) 'Glass in building – Part 1: Basic soda lime silicate glass products. Definitions and general physical and mechanical properties', BSI.

BS EN 1063 (2000) 'Glass in Building – Security glazing – Testing and classification of resistance against bullet attack', BSI.

BS EN 13501-2 (2007) 'Fire classification of construction products and building elements – Part 2: Classification using data from fire-resistance tests, excluding fire ventilation services', BSI.

For information on high-security glazing, see 'Protection against flying glass', http://www.mi5.gov.uk.

RECOMMENDED READINGS

Department for Communities and Local Government (1998) 'Approved Document N: Glazing – Safety in relation to impact, opening and cleaning', NBS.

Keiller, A., Walker, A., Ledbetter, S. & Wolmuth, W. (2005) 'Glazing at Height – Guidance for designers and clients', CIRIA.

20: Carpets
A short guide to floor coverings

'Do you want your underside felt, sir?' is a phrase you only hear in carpet shops. In fact, double entendres loom large where carpets are concerned. Whether it's running your fingers through soft downy piles, or having your weft warped, remember, when it comes to a good shag, sisal isn't important.

Marx may have developed the theory of Base and Superstructure but, not to be outdone, NBS 'General Guidance on floor coverings' clarifies the distinction between 'base' and 'substrate'. Using the definition in BS 8203 'Installation of resilient floor coverings', a base is the 'supporting structure to which the floor covering is to be applied' whereas a substrate is the 'surface of that base, or underlay, on which the floor covering is laid'. Confusion is introduced by BS ISO 2424: 2007 'Textile floor coverings – Vocabulary'. It defines substrate as 'construction, integral with the use-surface ... part of a textile floor covering directly exposed to traffic ... and composed of one or more layers, which serves as a support for the use-surface'. Maybe that definition lost a little in the translation, but essentially, a well-laid base is an important ingredient for well-laid floor coverings.

CONCRETE AND SCREED BASES

Coverings must not be laid on concrete bases that have a relative humidity above 75 per cent and only when they afford protection from ground moisture or water vapour. Bear in mind that a 50 mm thick screed can take up to two months to dry sufficiently and a 150 mm thick base, drying from one side, can take more than a year!

The warp is the yarn that runs lengthways, while the weft runs at right angles.

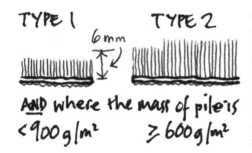

Where the relative humidity is too high, or there is insufficient programme construction time to allow adequate drying, a surface-applied damp-proof membrane can be considered to speed things up. The Contract Flooring Association's (CFA) 'Guide to Contract Flooring' states that surface damp-proof membranes, 'should always be considered a last resort and are clearly second best to correctly placed sandwich membranes and adequate drying times'.

Bases that have been chemically hardened or treated with a resinous seal may deleteriously affect flooring adhesives. However, BS 5442-1 'Classification of adhesives for use with flooring materials' was withdrawn on 1 Nov 2005 due to lack of interest from adhesives and sealant manufacturers.

TIMBER BASES

These must be level, rigid, sound and dry and have reached a steady moisture content equivalent to that which will pertain after the covering is laid. In-situ-applied wood preservative and flame-retardant chemicals can adversely affect flooring adhesives and cause severe and permanent distortion of textile floor coverings; such bases will require remedial treatment.

A fabricated underlay is essential where carpet is to be laid on plain or tongued and grooved timber boards, oriented strand board (OSB) supported on timber joists or generally on any particleboard substrates with gaps between adjacent boards in excess of 1 mm. Wood blocks may be a suitable base provided the blocks are clean, sound, firmly bonded and protected against moisture.

CLASSIFICATION AND MANUFACTURE

Carpets are designated Type 1 or Type 2 as shown in the drawing, although there are other significant and complicated classifications to provide a method of selecting suitable wearing capabilities with reliable appearance and performance criteria. The different levels of wear and appearance retention for different uses of carpets and carpet tiles are described in some detail in BS EN 1307 'Textile floor coverings – classification of pile carpet'. This relates specifically to machine-made pile carpets (and tiles) defined in BS 5557. Needled/fibre bonded carpets and tiles are covered in BS EN 1470 and BS EN 13297.

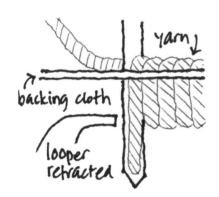

In a minority of cases, BS EN 1307 provides incorrect classifications for machine-made pile carpets by ignoring the inherent inaccuracies in their production methods and the variability of the appearance assessment. The BS EN 1307 classification system is widely accepted for commercial installations, but the Carpet Foundation has a domestic carpet scheme known as Quality Mark which lists a confirmed suitability classification for those that have undergone the flammability tests.

According to BS ISO 2424, tufts are the length of yarn, 'the leg or legs of which form the pile of a carpet'. In general, tufted carpets are created on a pre-woven backing fabric. Here, the needle carrying the yarn pierces this backing fabric pushing the yarn through to the underside. As the needle retracts, a hook or 'looper' captures the yarn, preventing it being pulled back out with the needle and creating a loop on one side of the backing layer. An adhesive or latex backing layer is applied to bind the tufts. It is the setting of the looper at the time of manufacture that determines the length of the pile. Loop piles – so called because the wearing surface is made up of uncut loops of yarn – produce a hardwearing, textured surface. To create a cut pile, a blade follows the action of the looper cutting through the loops and leaving individual strands.

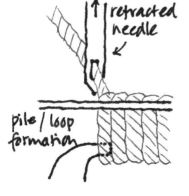

A tufting machine has a row of needles mounted in a needlebar running the width of the machine so that a full row of tufts is stitched in one cycle. The 'gauge' of the machine describes the needle density; for example, a 1/10 inch gauge has 10 needles to the inch.

Some other examples of tufted carpets include:

- Velvet carpet, otherwise known as 'plush' or 'velour' carpet, is a cut-pile carpet using yarns with very little twist so that the finished surface has no tuft definition.

- Shag pile rugs are quite different from other rugs in that the cut pile has greater than normal pile length and spacing between tufts. They tend to be fairly heavyweight and are usually manufactured in India, China, the USA, Morocco, Portugal or Israel.

Woven carpets such as Wilton and Axminster are also created on a primary backing fabric but create the pile bonding structure as part of the weaving process. The warp is the yarn that runs lengthways, while the weft runs at right angles. The selvedge is the interlacing of the warp and weft threads at the edge of a woven textile carpet to prevent it unravelling or fraying.

CONTRACT CARPETS

The most commonly used fibres for contract carpets are:

- polyamide (nylon): extremely durable
- polypropylene: very durable but lacks resilience
- pure new wool: durable and resilient.

Blended fibres are attempts to mix the best qualities and cost of each of the individual materials. For example, an 80/20 (80 per cent wool/ to 20 per cent polyamide) builds in the polyamide's durability and antistatic properties while retaining the feel of the wool. The CFA states that the relative lack of resilience in polypropylene 'is of no consequence in a flat fibre-bonded carpet, but when blended in up to 50 per cent by weight with wool in a tufted hard-twisted carpet, it adds bulk, stain resistance and significant cost reduction against an assessed acceptable reduction of the plus qualities of wool'.

BS EN 1470 and BS EN 13297 jointly cover all types of needled coverings and, through a common scheme, establish levels of use classes which indicate suitable use areas. These are described in the table below.

RESILIENT FLOOR COVERINGS

These are normally sheet floor coverings falling loosely into four categories: plastics, cork, linoleum and rubber. If used in conjunction with underfloor heating (which tends to operate at temperatures around 27°C), or where adhered floors are laid in locations with regular and extensive warming by direct sunlight, the adhesive may fail, and manufacturers' guidance should be sought. The response time of underfloor heating systems will be slower due to the insulating qualities of resilient floor coverings. All embedded heating pipes should be covered by a minimum of 25 mm screed.

Although Building Regulations Approved Document Part B doesn't stipulate the fire rating of floor coverings, it is advisable to use non-hazardous materials, especially on primary and escape routes.

> *The response time of underfloor heating systems will be slower due to the insulating qualities of resilient floor coverings.*

USE INTENSITY	DOMESTIC		COMMERCIAL		LIGHT INDUSTRIAL	
	CLASS	LOCATION	CLASS	LOCATION	CLASS	LOCATION
Moderate/light	21	Bedrooms	–		–	
Moderate	–		31	Hotels bedrooms, conference rooms, offices	41	Electronic assembly Precision engineering
General/medium	22	Living rooms, entrance halls	–		–	
General	22+	As 22 plus dining rooms and corridors	32	classrooms, small offices, hotels boutiques	42	storage rooms electronic assembly
Heavy	23	As 22+	33	corridors, stores lobbies, schools, open plan offices	43	storage rooms production halls
Very heavy	–		34	Multi-purpose halls, counter halls, department stores	–	

The following organisations provide guidance on specifying and laying carpets:

The Woolmark Company
www.woolfurnishings.com

British Carpet Technical Centre and Cleaning and Maintenance Research and Services Organisation (BCTC CAMRASO) www.bttg.co.uk

Contract Flooring Association (CFA)
www.cfa.org.uk

The Carpet Foundation
www.comebacktocarpet.com

Woolsafe www.woolsafe.org

Floor coverings produced from natural fibres such as sisal, seagrass, coir and hessian can be durable but are sometimes difficult to clean. Cork, the bark of the evergreen oak grown in Mediterranean countries, is a renewable material, and agglomerated cork floor coverings are manufactured from cork granules, a waste product of the bottle stopper industry. Traditionally the glue used to bind the granules together was ureaformaldehyde-based; nowadays many manufacturers use natural protein binders.

More detail on specifying and laying resilient flooring will be covered in a later Shortcut. Until then, the first rule of carpet care is to resist fluffing on the living room carpet.

REMEDIAL TREATMENTS FOR CARPET STAIN REMOVAL

stain → treatment ↓	artificial colours herbal tea	bleach	oil grease wax shoe polish	vomit	red wine		emulsion paint oil paint	coffee fruit juice	bubble-gum	nail varnish	ball point pen
1st attempt	1	1	2	3	4	5	1	6	7	8	9
2nd attempt	6	10	3	6	6		3				
3rd attempt				11							

KEY:
1 - COLD WATER ; 2 - 'WOOLSAFE - APPROVED' SPOT REMOVER ; 3 - 'WOOLSAFE-APPROVED' CARPET SHAMPOO ; 4 - BLOT WITH TISSUES;
5 - WHITE SPIRIT ; 6 - 'WOOLSAFE-APPROVED' WATER-BASED SPOT REMOVER; 7 - SOLVENT REMOVER ; 8 - ACETONE ; 9 - SURGICAL SPIRIT;
10 - RE-COLOURING KIT ; 11 - 'WOOLSAFE - APPROVED' DISINFECTANT / DEODORISER

References

BS 8203 (2001) 'Code of practice for installation of resilient floor coverings', BSI.

BS ISO 2424 (2007) 'Textile floor coverings – Vocabulary', BSI.

Jacobs, B. (1968) 'The History of British Carpets', CFR.

Ward, D. (1969) 'Tufting', Textile Business Press.

RECOMMENDED READINGS
BS EN 1307 (2008) 'Textile floor coverings – classification of pile carpet', BSI.

Contract Flooring Association (2007) 'The CFA Guide to Contract Flooring', CFA.

Crawshaw, G.H. (2002) 'Carpet Manufacture', CRC.

Robinson, G. (1972) 'Carpets', Textile Book Service.

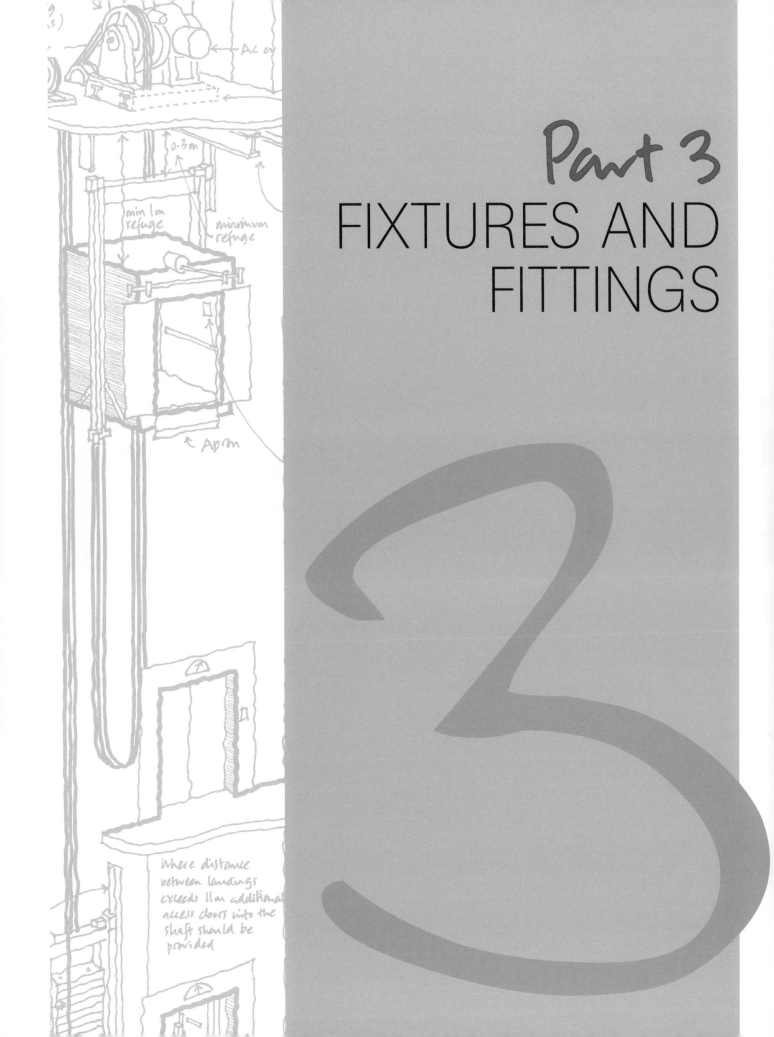

Part 3
FIXTURES AND FITTINGS

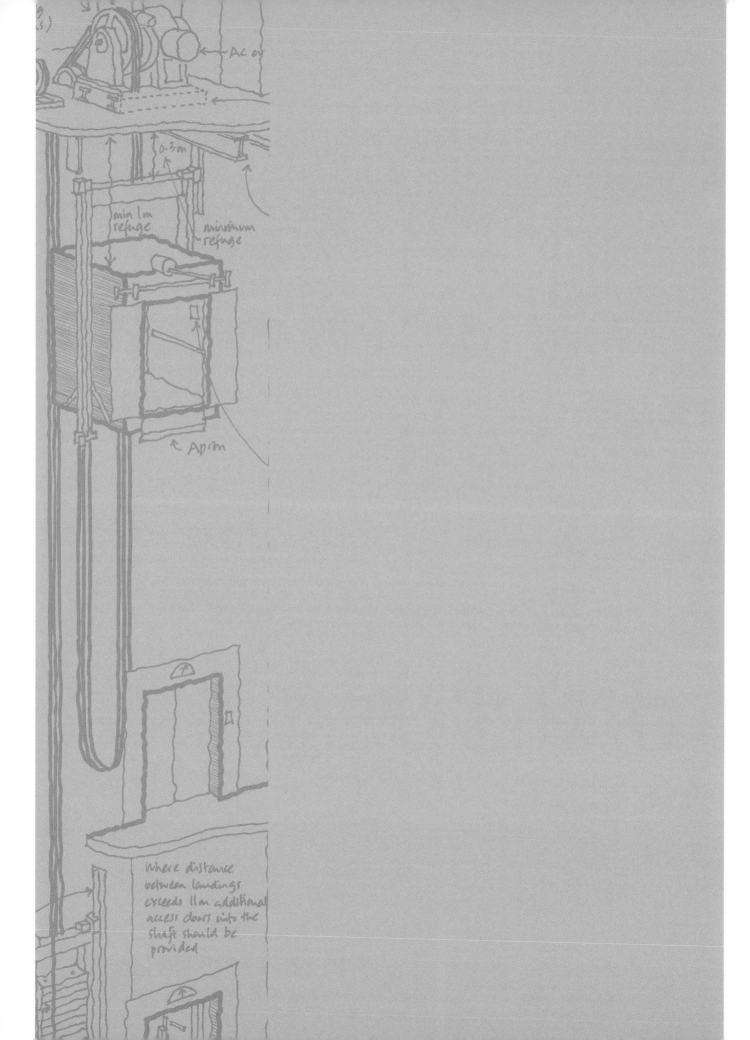

Ac o←

0.3m

min 1m
refuge

minimum
refuge

← Apron

Where distance
between landings
exceeds 11m additional
access doors into the
shaft should be
provided

21: How Locks Work
Key facts about how to obtain closure

Mortice, rim and deadbolt isn't a firm of solicitors, but just some of the outsourced specification data that architects regularly include in their ironmongery schedules without giving it a second thought. Three lever or five lever? Pin tumbler versus cylinder latch? Lever handle or the cheeky knobset? What do they mean and how do they work?

Whichever crime statistics you choose to believe, the chances of an average household being burgled are around once every 35–50 years. Not bad odds, but still we try to devise more and more ingenious ways to keep over-hyped intruders out of our houses.

Actually, domestic front door locks haven't changed much for 150 years; literally true if you care to examine the Easter Island display at the Science Museum's magnificent lock exhibition. Even in the West, the materials are different and their appearance is more varied, but locks' essential operation is unchanged. Yale even promotes itself as emulating the traditions of Egyptian locks and keys – dating back to Nineveh and its palaces, 4000 years ago.

Mind you, security consciousness has its downsides. In the 16th century, Ivan Vasilyevich, Tsar of all Russia, locked his wife in her room while he went off to war. To make sure no one could use a duplicate key to 'gain entry' he had the locksmith beheaded. Tragically she fell ill and, unable to get medical help, she died – reputedly turning a relatively benign ruler into the legendary Ivan the Terrible.

A deadlock is a mortice lock that operates by key only, whereas a sashlock combines a deadlock with a separate spring-loaded latch.

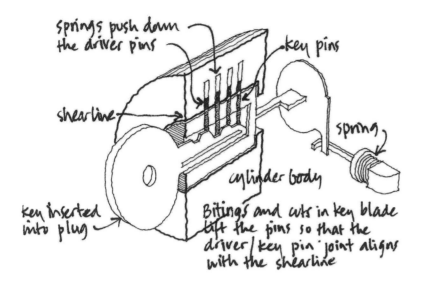

springs push down the driver pins

key pins

shearline

spring

cylinder body

key inserted into plug

Bitings and cuts in key blade lift the pins so that the driver/key pin joint aligns with the shearline

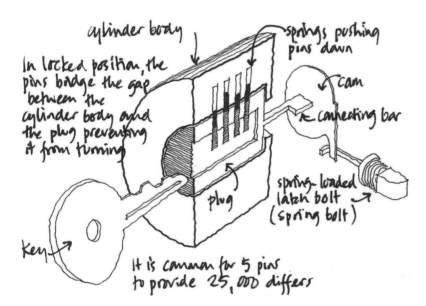

cylinder body

springs pushing pins down

In locked position, the pins bridge the gap between the cylinder body and the plug preventing it from turning

cam

connecting bar

plug

spring-loaded latch bolt (spring bolt)

Key

It is common for 5 pins to provide 25,000 differs

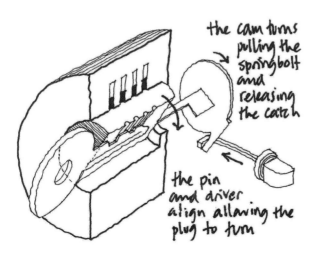

the cam turns pulling the springbolt and releasing the catch

the pin and driver align allowing the plug to turn

At the turn of the 19th century, British locksmiths Robert Barron and Joseph Brammah developed the idea of having a small key that didn't reach the bolt but acted through intervening moving parts. In these, the lock turns a cam that pulls a latch back, allowing the door to open. A spring usually pushes the bolt back out again. A *non-spring-loaded* latch called a deadbolt (requiring a key to open and close the bolt) is usually more secure as it is more difficult to push the bolt in from the side of the door.

Jeremiah Chubb of Portsmouth patented the detector lock in 1818, which won a government challenge for a lock that could not be opened by any other than its own key. Even a professional locksmith who was imprisoned on a ship in Portsmouth docks – with the incentive of a promise of a pardon and £100 – failed to open it after trying for three months.

The cylinder pin tumbler lock, by Linus Yale of New York in 1848, uses Barron's double tumbler principle invented 70 years earlier. It has rows of two sets of pins, one above the other – the 'double tumblers' – which stop the cylinder from moving. The correct key raises the pins such that the joint lines between them line up with the cylinder edge. This frees the cylinder to move within the barrel, allowing it to move and turn the cam. This cam moves the locking bolt – or latch – in and out, and, with minor modifications, this remains the basis of Yale locks today (see diagram, which also lists some of the definitions commonly associated with locking devices).

EXTERNAL DOOR SECURITY

A mortice lock (the name is taken from mortice and tenon joints in woodworking) is one that shoots a deadbolt into a mortice, or 'keep', in the doorframe. A deadlock is a mortice lock that operates by key only, whereas a sashlock combines a deadlock with a separate spring-loaded latch, the latter operated by a doorhandle enabling the door to be secured without always having to use a key.

The majority of locks fitted to external doors in the UK are to BS 3621: 2004 simply because most British insurance companies insist on it. In short, this standard requires that the locks be provided with: at least five levers, anti-pick notching, locking devices separated from any handle operation, minimum 14 mm bolt projection, anti-drill plates, concealed screw fixings and the locking mechanism manufacture must be one of 1000 keying variations. (Note: A three-lever lock is unlikely to have more than 100 'differs', as they're called).

Over 50 years ago a common Ingersoll lock boasted ten levers and, until recently, Chubb commonly manufactured locks with seven levers, but discontinued them in 2004 because the more levers, the thinner they have to be in order to fit in the lock case. Today, no mainstream manufacturer produces locks with seven levers.

Levers and lever handles are different things. Levers – as in 'five-lever latch' – are the devices within the casing that lock the bolt in place and need to be raised to allow the deadbolt to move forwards and backwards. In its simplest form, this is done by the indentations (step heights) of the key bit, i.e. the uneven cuts in the section of the key that is inserted in the keyhole. Turning the key lifts the levers differentially, depending on their pattern, in order to align a slot, or 'gate' in each of the brass levers so that an obstruction, or 'stump' can pass through and the bolt travel forward (see diagram). Each lever is spring-loaded so that the gate shuts after the stump has passed capturing it on the other side of the lever and maintaining the bolt in its extended position. The reverse process is needed to open the gate and retract the deadbolt. Notches are regularly cut into the lever to try to fool lockpickers about the location of the gate.

It is questionable how many of today's young opportunist burglars have the professional integrity to carefully pick a lock to break in. Insurers will undoubtedly tick off kitemarked, British Standard 'thief resistant' locks but may reject your claim because of your flimsy doorframes or large glass panes. Sturdy specification is the first defence against breaking (rather than lock-picking) and entering and Secure by Design criteria often request additional safety features such as anti-jemmy protection, steel-faced solid core doors, restrictors on letter plates, and external grillage, depending on the level of security required.

The Door and Hardware Federation (DHF) 'Best Practice Guide' on mechanically operated locks highlights the 11 classifications (taken from BS EN 12209: 2003) to be graphically marked on doors identifying everything from the door mass to the ironmongery's corrosion

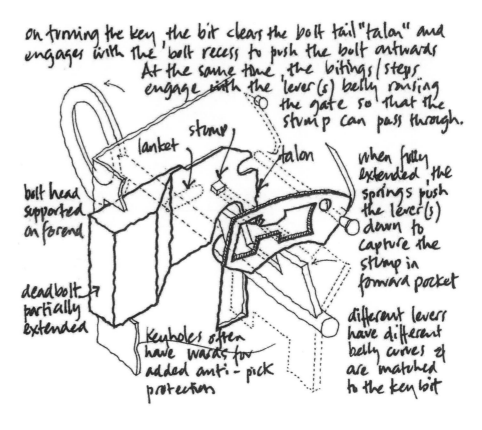

lanket groove/hole to aid the bolt's movement

cutaways to avoid cap screws

key inserted and turned to engage the lever & bolt

stump

stump held in pocket of lever(s) preventing forward motion

levers spring-loaded

pivot

lever(s) with different curved bellies to match key

deadbolt retracted

keyhole

Key bit

forend

Key biting / key step

cutaway of lockcase

on turning the key, the bit clears the bolt tail "talon" and engages with the bolt recess to push the bolt outwards. At the same time, the bitings / steps engage with the lever(s) belly raising the gate so that the stump can pass through.

lanket

stump

talon

bolt head supported on forend

when fully extended, the springs push the lever(s) down to capture the stump in forward pocket

deadbolt partially extended

Keyholes often have wards for added anti-pick protection

different levers have different belly curves & are matched to the key bit resistance. The seventh digit (or seventh DHF icon – indicated by a silhouetted padlock) lists seven grades of security and drill resistance: from Grade 1 'minimum security and no drill resistance' to Grade 7 'very high security with drill resistance'. The ninth DHF icon identifies the keying operation (grade 0 has no locking; grade H is automatically locked) and the tenth icon identifies the spindle operation. (The Guild of Architectural Ironmongers technical manual 'Locks and latches – Part 1', written a few years before BS EN 12209, contains only the first seven digit/icon classification system.)

Locks that operate with keys from both sides have a symmetrical arrangement of levers meaning that the number of differs reduces dramatically, effectively any lockpicker need only identify the first three lever gate levels and the remaining two are the mirror image. Therefore, it should be realised that not all five-lever locks provide the necessary 1000 differs to show compliance with BS 3621. Additional protection is normally afforded by a 'ward' and/or 'bolt thrower' which shrouds the levers from lockpicking tools as well as pushing the bolt without direct contact from the key bit. Cylinder locks perform a similar security function with the added bonus that cylinders can be replaced without having to change the entire lock case.

Returning to the opening paragraph, the risk of burglary is fairly remote. Statistically, at least, if you left your doors open you'll probably find that no one has bothered to take advantage.

Note: Definitions are taken from BS EN 12209: 2003 (which supersedes BS 5872: 1980).

Thanks to the Guild of Architectural Ironmongery and particularly to Master Locksmith, Mr Lewis Beadle of Abilock Lock Technicians.

> "Levers – as in 'five-lever latch' – are the devices within the casing that lock the bolt in place and need to be raised to allow the deadbolt to move forwards and backwards.

References

BS 3621 (2007) *'Thief resistant lock assemblies – Key egress'*, BSI.

BS EN 12209 (2003) *'Building hardware – Locks and latches – Mechanically operated locks, latches and locking plates – Requirements and test methods, (AMD Corrigendum 16436)*, BSI.

Defence Estates (1998) Historic Buildings Factsheet T 7.03: *'Metalwork – external elements: ironmongery and special fittings'*, DE.

Door and Hardware Federation (2006) *'Cylinders for locks to BS EN 1303: 2005. Best practice guide'*, DHF.

Door and Hardware Federation (2005) *'Mechanically operated locks, latches and locking plates to BS EN 12209: 2003. Best practice guide'*, DHF.

RECOMMENDED READINGS

Door and Hardware Federation (2006) *'BS 3621 and BS 8621: 2004 Thief-resistant lock assemblies. Best practice guide'*, DHF.

Guild of Architectural Ironmongers (2001) Technical Manual 1.4: *'Locks and latches – Part 1'*, GAI.

Guild of Architectural Ironmongers (2000) *'Locks and latches – Part 2'*, GAI.

22: Fixings and Fasteners
Nails and screws

For some people, it's a toss-up over the choice of a screw or an annular ring shank. But actually, each performs distinct holding down functions. Ideally, different nails and screws should be specified for different materials and situations. Here we examine the choices available.

When is a fixing not a fixing? When it's a fastener? Well, not really, sometimes they can be one and the same thing. The best definitions are as follows:

- 'Fixing' is the act of holding and securing an object in place (sometimes called the fixing method); and

- 'Fastener' is the holding down and securing connectors used for fixing (sometimes called the fixing device).

The confusion arises because 'fixing' is a gerund – that is, it can act as a verb or a noun – and so it helps to insist on its use as a verb only, where possible. Once it becomes used interchangeably, as it invariably is, the meaning can become confused. Using the word 'fastener' is a way of clarifying this difference.

> When using a nail gun, be aware that firing the nails into unsupported 6 mm plywood, say, or into any other thin material that is not directly located over a timber stud or similar substrate, will result in the nail being shot through the material like a bullet.

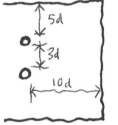

hardwood   softwood

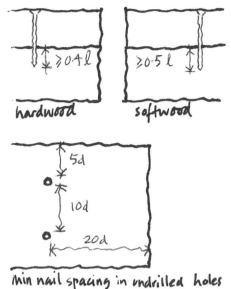

min nail spacing in undrilled holes

min nail and screw spacings in drilled holes

With this differentiation in mind, 'fixings' describes a system for holding and securing components together or into place. For example, where a joist is nailed into a joist hanger with additional lateral restraint straps, the 'fixings' represents the complete system of straps, connectors, etc., whereas the 'fasteners' would be the nails. In the event of simply fixing (verb), where maybe a rafter is skew nailed to a wall plate, then 'the fastener' is the 'fixing' (noun), but as suggested above, it would be clearer to specify the fastener as the nail and the 'fixing' as the nail centres. That's cleared that up. Except to say that the threaded part of a screw fastener is actually known as the fastening...

Anyway, this Shortcut provides an introduction to the various types, functions and uses of some of the huge range of nails and screw fixings. It will focus on common scenarios (wood screws, for example) and common tasks to which they are best suited.

NAILS
Wire nails date from the late 19th century. Before that, 'cut nails' were common, punched out, or guillotined, from a flat plate of rolled iron (see 'floor brads' and 'cut clasp nails' in the box).

Nails, staples, wood screws, coach screws and bolts are all variants of dowel fasteners. The Timber Research and Development Association (TRADA) points out that all of them may be used for laterally loaded connections, but for axially loaded connections, only nails, screws and bolts are normally used. As such, nails are normally required to resist just the shear forces at the interface (or possible interstice) between two or more joined materials.

Eurocode 5 'Design of timber structures', which outlines the common rules for structural timber use in buildings, specifies a minimum tensile strength and provides design procedures for all of these fasteners. It will replace BS 5268 in about 2010, but until then BS 5268-2 refer to compliance standards for the fasteners that it covers.

Joining two pieces of hardwood, at least 40 per cent of the nail length must extend into the lower piece; joining softwoods, that should increase to at least 50 per cent. When hammering nails with a diameter greater than 5 mm into hardwoods, pre-drilled holes should be used to avoid splitting the timber. Similarly, for nails more than 100 mm long, pre-drilling is recommended. To ensure a good grip, the hole should be no more than 80 per cent of the nail diameter. Pre-drilled holes should be at least 10 times their diameter from the end grain of hardwood; where there are no pre-drilled holes, nails should be positioned at least 20 times their diameter from the end grain.

When using a nail gun, be aware that firing the nails into unsupported 6 mm plywood, say, or into any other thin material that is not directly located over a timber stud or similar substrate, will result in the nail being shot through the material with the velocity of a bullet.

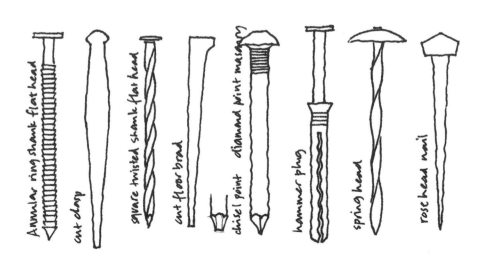

NAIL AND SCREW USES

Some common uses of screws and nails are as follows:

Countersunk screws – Primarily for fixing timber to timber, or metal to timber, into a pre-drilled hole so that the top of the flat screw head sits flush on the surface. These screws' shanks have a range of diameters from 4 to 8 mm.

Round head screws – For metal to wood connections where there is no countersink. The range of diameters is as above.

Raised head screws – For quality finished items that may have to be removed on regular basis: mirrors, glazed panels, etc.

Snap-off screws – A segmented thread (gaps on thread at 5 mm intervals along the lower part of the shank) to enable reduction in length to suit application, e.g. through door ironmongery fixing (letterboxes, say).

Clutch head screws – A flush head screw with slot designed to prevent it being unscrewed.

Screw cups – A pilot hole (rather than a countersunk clearance hole) is needed to accommodate the screw cup metal collar which sits flush with the surface and effectively increases the bearing strength under the screw head.

Dome head screws – A decorative dome head finish is screwed into a threaded hole in the top of a wood screw: ideal for mirrors, etc.

Coach screws – Heavy-duty fixing screws to timber or plugged masonry with exposed hexagonal (or similar) head which is turned by a spanner. These have a range of diameters from 6 mm to 20 mm and lengths of 25 mm to 300 mm. They are regularly used in engineered timber structures.

Self-tapping screws – These are often used for fixing relatively thin sheet metal material although can equally be used with plastics, plywood, etc. and come as pointed (Type C) and blunt (Type F) previously known as AB and B respectively. They are called self-tapping screws because they act as a screw auger as they are being screwed into a material, automatically creating a female thread with which to bind. The gauge range for Type C is between 4–14 g; and for Type F is 2–14 g.

Wire nails – These are usually round in section and made from mild steel. The term refers to the method of manufacture, and means any nail made from a cut wire coil with a minimum tensile strength of 600 N/mm². It has a range of diameters from 2.65 mm to 8 mm. Bright or blue nails have no finish and will corrode easily in external conditions. Wire nails are distinct from 'forged nails' or 'cut nails' (nails cut from a plate).

Clout nails – Commonly 25 mm long, these are regularly used for fixing laths, insulation board, partitions and ceilings.

Lost head round wire nails – Commonly used for timber to timber fixing (fascia boards, weatherboarding, etc.). They resist timber deformation well and are usually punched below surface and easily hidden.

Annular ring shank nails – Serrated shank nails used for fixing, for example, plywood flat roof deck owing to their excellent grip (or what TRADA calls 'withdrawal strength').

Square twisted shank flat head nails – Between 40–65 mm long, these are regularly used with hardwoods and have excellent pull-out strength and are regularly used for fixing joist hangers and restraint straps.

Convex head roofing nails – These have either a chisel or, more commonly, a diamond point (i.e. a four-sided pyramid). The chisel point should be inserted across the grain so as not to split the timber.

Spring head nails – Galvanised twisted shank nails with a large 'umbrella' head, designed to fix boards and sheet materials.

Cut clasp nails – These are flat nails with a punch head, widening out in the middle of the shank. They are often used for rough fixing timber to blockwork.

Cut floor brads – A flat nail with one vertical side and one angled side and an eccentric head. Used primarily for fixing floorboards.

Masonry nails – Hardened steel nails, from 12 mm up to 85 mm, for driving into solid bricks or blockwork with a heavy hammer. The smallest are known as masonry pins and are used to hold thin sheet materials in place. The minimum penetration into the masonry should be at least 25 mm (excluding the thickness of any plaster).

Rose head flat points – Often used for conservation work, the shaft has a long taper in one direction and pinched in just below the head. For use on green timber which will expand around the nail after fixing.

Hammer plugs – As the name suggests, this is a plug with a nail inset which is inserted into a pre-drilled hole. When the nail is hammered home, the plug opens out to bind with the sides of the hole for excellent pull-out strength.

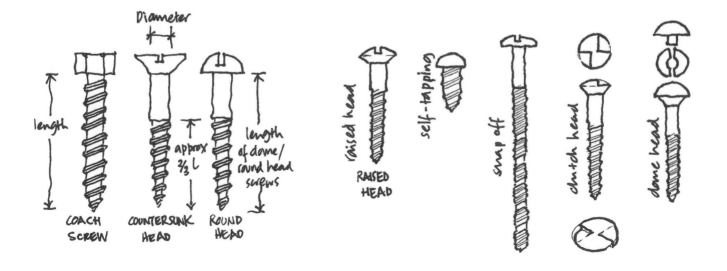

SCREW GAUGE

Screws are specified by gauge (determined by the major diameter, usually across the non-threaded portion of the shank) and length (from the top of the head to the gimlet point), for example, No. 8 x 25 mm. This does not include the extra height of round- or dome-headed screws.

To check whether two screws are the same, hold them next to one another with the heads at opposite ends and if their threads mesh together then they have the same pitch.

The relationship of gauge and diameter is shown below:

GAUGE	MAJOR DIAMETER
6 g	3.5 mm
8 g	4.2 mm
10 g	4.8 mm
12 g	5.5 mm
14 g	6.3 mm

Screws with a greater diameter than 5 mm should be turned in pre-drilled holes to prevent splitting the wood.

When driving brass or aluminium wood screws into hardwoods the torque required can often exceed the shear resistance of the metal. Consequently the head may snap off. Therefore, it is advisable to use a steel screw first to cut the thread.

BS 1580-1: 2007 describes 'Unified screw threads' as follows:

UNC: Unified coarse thread is for 0.25 inches to 4 inches diameter

UNF: Unified fine thread 0.25 inches to 1.5 inches diameter

UNEF: Unified extra fine thread 0.25 inches to 1.6875 inches diameter

SCREWS

The Construction Information Service suggests that BS 1210: 1963, 'Specification for Wood screws' is 'obsolescent but still relevant', predominantly because many of the referenced British Standards are obsolete. Even though the dimensions in the standard are written in inches, this is not enough to condemn it, being just one more of those British construction industry curiosities which says that certain things are legitimately to be measured in imperial units. But even though the Standard has now been in force for 45 years, there is still an uncorrected howler in Annex G: where the final calculation factor has been written t_1/t_2 instead of t_2/t_1! Fortunately this has little significance for designers. To comply with the standard, all screws must have a minimum tensile strength of 550 N/mm^2 – excluding coach screws – and have a shank diameter greater than 10 mm. In order to address some of these issues, a new suite of British Standards for screws (BS 1580: 2007) has just been released, although there is still no European Standard.

BOLTS

For bolted connections, washers must have a minimum external diameter and thickness, 3 times the diameter of the bolt (the hole itself may be up to 2 mm wider than the bolt). TRADA recommends that the thickness of washer must be 0.25 times (in 'Wood Information', Section 2/3 Sheet 36: 2003) but 0.3 times the diameter (in 'Wood Information', Section 2/3 Sheet 52: 2002).

PERFORMANCE

When comparing the jointing performance of fasteners of equivalent diameters:

- Nails generally have the advantage in terms of lateral load-carrying capacity.

- Screws have better axial withdrawal resistance.

- Dowel joints, in the form of timber dowels or through bolts, provide significantly higher load carrying capacity.

References

BS 919-4 (2007) 'Screw gauge limits and tolerances. Limits of size for gauges for screw threads of unified form diameters ¼ in and larger', BSI.

BS 1202-1 (2002) 'Specification for nails. Part 1: Steel nails', BSI.

BS 1210 (1963) 'Specification for Wood screws' (deemed obsolescent but still relevant), BSI.

BS EN 383 (1993) 'Timber structures. Test methods. Determination of embedding strength and foundation values for dowel type fasteners', BSI.

BS EN 409 (1993) 'Timber structures. Test methods. Determination of the yield moment of dowel type fasteners. Nails', BSI.

BS EN 845-1 (2003) 'Specification for ancillary components for masonry. Ties, tension straps, hangers and brackets', BSI.

BS EN 10230-1 (2000) 'Steel wire nails. Loose nails for general applications', BSI.

BS EN 1995-1-2 (2004) 'Eurocode 5: Design of timber structures. General – Structural fire design (AMD Corrigendum 16498)', BSI.

DIN 571 (1986) 'Hexagon head wood screws', Deutshces Institut für Normung.

RECOMMENDED READINGS

BS 1580-1 (2007) 'Unified screw threads. Screw threads with diameters ¼in and larger. Requirements', BSI.

BS 5268-2 (2002) 'The structural use of timber. Permissible stress design, materials and workmanship', BSI.

BS EN 1995-1-1 (2004) 'Eurocode 5: Design of timber structures. General – Common rules and rules for buildings (AMD Corrigendum 16499)', BSI.

Office of the Deputy Prime Minister (2004) 'Approved Document A – Structure', ODPM.

23: Turning a Deaf Ear Assisted hearing devices for the hard of hearing

The placing and operation of acoustic loops and other systems in public buildings need careful consideration. Incorrect specification can result in an unsatisfactory audible environment for the hard of hearing as well as the possibility that electronic equipment, within range of the acoustic loop's magnetic field, may be adversely affected.

In 2005, a design collaboration between the Royal National Institute for Deaf People (RNID), Blueprint magazine, and creative agency Wolff Olins, was set up to try to convince the public that hearing aids, rebranded as 'hearwear', should become as stylish as designer glasses. By making people realise that ungainly, badly flesh-toned plastic earpieces are a thing of the past, the designers hope to encourage some of the 9 million UK residents who are hard of hearing to use hearing aids ('hard of hearing' includes those with 'some useful hearing' (see levels of deafness classification at the end of this Shortcut).

Recent surveys show that only 2 million UK residents own hearing aids and only 1.4 million *admit* to using them regularly. There is still, it seems, something of a stigma associated with hearing loss. A bit like bifocals, they tend to suggest that you are getting on a bit and any help in encouraging people to be less abashed about their hearing loss is an admirable objective. The most recent data shows that there are just under 55,000 registered deaf people in England, an increase of 17,000 (or 45 per cent) since 1989, partly explained by the fact that the population is aging, but also by the improvements in the appearance and performance of the hearing aids themselves.

There has been a 45 per cent increase in registered deaf people in England since 1989, partly because the population is aging, but also because of the improvements in the appearance and performance of hearing aids.

Over the years, deaf and hard-of-hearing children have begun to be educated in mainstream – rather than 'special' – schools, and the British Association of Teachers of the Deaf (BATOD) has assessed that these pupils account for about 75% of all deaf children of school age. One of the simplest methods of enhancing their ability to hear the lesson clearly is to reduce background noise by siting classrooms away from playgrounds, busy circulation spaces and communal areas. Improving acoustic separation, and keeping classroom occupants quiet also helps.

Coincidentally, there is a growing health and safety campaign by the National Union of Teachers, backed by the Department for Education and Skills (DfES) pointing out that teachers are more likely than most other occupational groups to consult doctors about voice disorders. A 2007 study of around 4000 teachers in France discovered that they were twice as likely as other workers to suffer disorders ranging from sore throats to vocal fold swelling. American teachers take an average of two days sick leave per annum due to voice problems, reputed to cost £315 million in healthcare and replacement costs. As a solution, microphones, which have been used in schools in America since the early 1970s, may find their way into UK classrooms, especially since, according to the British Association of Audiological Scientists (subsequently merged and renamed the British Academy of Audiology), amplified sound combined with low background noise can have more of a beneficial impact on the hearing ability of hard-of-hearing students than fine-tuning the reverberation times of fittings and finishes.

THE SOUNDFIELD SYSTEM

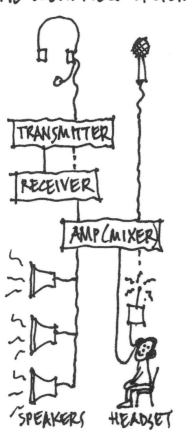

SPEAKERS   HEADSET

AMPLIFIED SOUND

An amplified audible sound system is an improvement on making hard-of-hearing children wear a headset in class (although this may still be necessary for pupils with significant hearing loss). The 'sound field system' comprises a teacher's headset and transmitter, an amplifier and a number of audio speakers to provide an equal distribution of sound whichever way the teachers (or pupils) are facing. The system uses a wireless link between the microphone and the amplifier, operating on VHF or UHF radio, or via an infrared link, and is relatively simple to set up, balance and modify.

INFRARED SYSTEMS

While a sound field system may be suitable for some school and lecture situations, it is not always desirable to improve audibility in public spaces by magnifying the sound for everyone. Infrared systems (available with a stereo facility) use invisible infrared light to transfer sound from the 'radiator' (a transmitter linked to a speaker's microphone and amplifier, for example) to receivers' headsets. It may be necessary to have several radiators depending on the size and shape of the venue. In general, infrared systems are directional, meaning that headset wearers may lose sound quality if they turn their heads. Care should also be taken to ensure that infrared-absorbent wall coverings and furnishings are not used.

AFILS

One of the more common – and cheapest – methods to improve the sound for targeted users is the use of audio frequency induction loop systems (AFILS). Approved Document M: 2004 'Access to and use of buildings' states that AFILS are needed in (somewhat ill-defined) 'reception points' of buildings other than dwellings. Designers should also note that badly lit reception points, or those enclosed by tinted glazed screens, can sometimes compromise the ability of deaf persons to lip read or follow sign language.

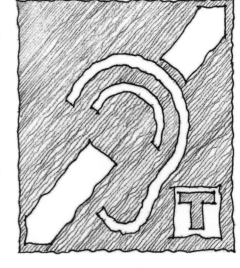

In 1985, the International Federation of Hard of Hearing People (IFHOH) adopted the induction loop icon as the official symbol for AFILS. These signs should identify the presence of the AFIL system as well as the prime locations for best reception. Ideally, approved signage should also be displayed where reception is *not* possible, with the icon showing that there is no signal. This avoids the hard of hearing wondering whether it is their hearing aid that is at fault.

An acoustic loop is simply a coil of electrical wire (preferably solid rather than stranded) that is connected in a loop to the output device (an amplifier linked to a microphone, radio, etc.). Essentially, an audio-frequency electric current circulates in the wire (varying with the speech, or music played through the amplifier); this, in turn, generates a magnetic field within the loop. The alternating magnetic field replicates the input sound modulation and this can be picked up within range by suitable receivers and reconverted back into sound. Receivers in hearing aids must be switched to 'T' (for 'telecoil') for the wearer to pick up the useful component of the magnetic field, and once tuned in, he or she can hear without having to rely on any visible paraphernalia drawing attention to their deafness. In normal situations, hearing aids are switched to 'M' for 'microphone' which allows the wearer to pick up normal close contact sounds; and for those with hearing aids without an M-T switch, a special adaptor is required to benefit from acoustic loops.

When setting out an induction loop circuit, consideration must be given to the listening plane (the plane which lies ideally at the median level of the ears of the listeners, around 1.2 m above finished floor level). Loops are often embedded in the walls at this height, but because the magnetic field gets stronger nearer to the loop, disturbances can be caused to hearing aid wearers if they are seated around the perimeter wall. Placing the loop in the slab, or a floor or ceiling void is acceptable although this requires more current for the same field strength at the hearing plane. Wherever the location, all systems must be satisfactorily designed to provide what BS 7594 calls 'a sufficiently (but not excessively) strong, and uniform useful component of magnetic field within the required working area'.

In a horizontal loop the magnetic field is in the vertical plane and lines up with the vertical magnetic pick-up coil in the hearing aids of typically sedentary audience. If the loop is in a hospital, where the hearing aid wearers may be lying down for considerable periods of time, the loop should be vertical as it is the perpendicular (horizontal) magnetic field component that is important. Similarly, in large auditoria, several loops will be required at

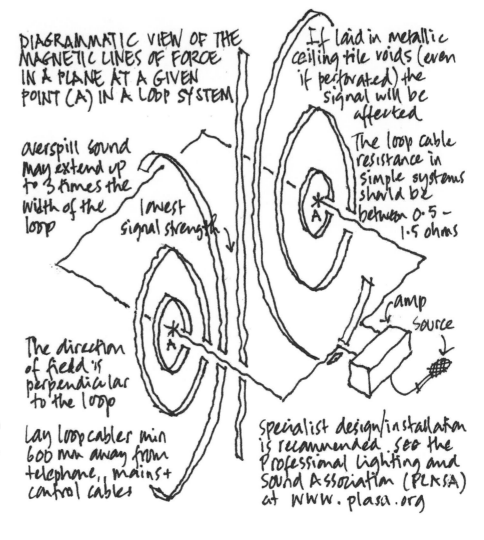

DIAGRAMMATIC VIEW OF THE MAGNETIC LINES OF FORCE IN A PLANE AT A GIVEN POINT (A) IN A LOOP SYSTEM

If laid in metallic ceiling tile voids (even if perforated) the signal will be affected

The loop cable resistance in simple systems should be between 0.5 – 1.5 ohms

overspill sound may extend up to 3 times the width of the loop

lowest signal strength

The direction of field is perpendicular to the loop

Lay loop cables min 600 mm away from telephone, mains + control cables

amp
source

specialist design/installation is recommended. See the Professional Lighting and Sound Association (PLASA) at www.plasa.org

LEVELS OF DEAFNESS:

It is important to understand the needs of the hard of hearing user group and to design accordingly.

NB: dBHL means 'decibel hearing level' and indicates a person's hearing relative to accepted standards for normal hearing. For example, 50 dbHL indicates a hearing loss of 50 db.

Mild hearing loss (25–40 dBHL) Difficulty in following speech.

Moderate hearing loss (41–70 dBHL) Difficulty in noisy situations; can use telephone with amplification.

Severe hearing loss (71–95 dHBL) Lip-reading is an essential skill. Difficulty hearing amplified telephones. Text messaging is increasingly used.

Profound hearing loss (96+ dHBL) Hearing aids are of little or no help. Lip-reading, signing and/or writing and texting may be the main communication tools.

various heights to ensure that each tier benefits, although there must be no magnetic interference between each separate loop. While the RNID suggests that in domestic conditions loops are not too badly affected by having to lay the perimeter wire around obstructions (doorways, windows, etc.), in larger applications the effect on the magnetic field caused by a non-uniform layout should be carefully calculated.

The system must be designed and installed for maximum efficiency, and BS 7594 states that it must not produce 'an unacceptably extended coverage area which could cause interference with other systems or compromise confidentiality'. Solid (not twisted core) wires with core diameters of 0.5–2.5 mm (complying with BS 6500 and chosen to suit the specifics of the room layout) can be housed in plastic tubing for added protection but must not be laid in the vicinity of other horizontally laid electrical wiring, as this will cause magnetic interference and consequently create an annoying background hum in the ears of hearing aid wearers. Similarly, if there are metal conductors in the vicinity of the loop (metal stud walls, steel columns, for example) this too will give rise to poor reception by weakening or blocking the signal.

Conversely, the magnetic field can spread beyond the designated room, meaning that those with telecoil functions on their hearing aids can pick up conversations in other rooms. This can be irritating for the wearer and also worrying for the speaker, especially in a home-based system where you expect private conversations to stay private. Often, multiple loop systems, laid into the floor slab or void are sufficient to combat this, but there are a number of proprietary systems that deal with this issue. Where speech is likely to be of a confidential nature, e.g. doctors' surgeries, an infrared system may be more appropriate.

Professional design guidance and installation, as well as thorough pre-Completion Certificate testing, are essential to minimise this problem. After installation, the system should be calibrated and its controls 'locked' to prevent unauthorised or accidental alterations. Maintenance readjustment, on the other hand, is seldom needed provided that the function of the room doesn't alter significantly, but it is wise to provide accessible junction boxes should it be necessary. On completion, the person responsible for the building should be given full instructions on how to use the system. Ensure, in the first instance, that it is turned on!

Finally, where there is no embedded loop technology, portable loops are handy for face-to-face contact and also help to minimise the spread of the magnetic field. These loops comprise a briefcase-sized unit that are usually adequate for discreet conversations within a 1 m radius.

References

BS 6500 (2000) 'Electric cables – Flexible cords rated up to 300/500 V, for use with appliances and equipment intended for domestic, office and similar environments' (AMD 13631) (AMD 14200) (AMD Corrigendum 15407) (AMD 15651) (AMD Corrigendum 16644), BSI.

BS 7594 (1993) 'Code of practice for audio-frequency induction-loop systems (AFILS)', BSI.

Office of the Deputy Prime Minister (2004) 'Approved Document M: Access to and use of buildings', ODPM.

RECOMMENDED READINGS
British Association of Teachers of the Deaf (2001) 'Classroom Acoustics – Recommended Standards', DfES.

BS EN 60118 (2006) Part 4: 'Electroacoustics. Hearing aids. Induction loop systems for hearing aid purposes. Magnetic field strength', BSI.

National Health Service (2006) SDA910, The Information Centre for health and social care, Social Care Statistics 'Information and Guidance for the Registered Deaf or Hard of Hearing collection', NHS.

The Stationery Office 'Disability Discrimination Act 1995', TSO.

24: How Escalators Work
Rolling out moving pavements

There are estimated to be 7100 installed escalators in the UK, supplied and maintained by what the Office of Fair Trading calls 'the big four' (i.e. Kone, Thyssen, Schindler and Otis). But how do escalators and travelators work, and what are the issues for specifiers? This Shortcut looks at the relatively slow movement of escalator technology.

At a Rem Koolhaas-designed visitor centre in Essen, Germany, a subsidiary of ThyssenKrupp Elevator (with the unfortunate Carry-On-esque name ThyssenKrupp Fahrtreppen) claims to have installed the longest escalators in the world. Located in Zollverein, the former colliery-turned-UNESCO World Heritage Centre, the escalators are around 60 m long and rise 23 m vertically. But don't always believe the hype: London's Angel tube station's escalator is longer. Even the Tyne Pedestrian Tunnel (installed in 1951 and still working) is 61 m long, with a 25.8 m vertical rise.

In his Seattle Public Library scheme, opened in 2004, Koolhaas specified that the escalators be 'bright chartreuse'; at Zollverein, they are merely bright orange. But a lick of paint and an increase in length doesn't alter the fact that the escalator hasn't really had any major changes to its form and function since Otis got together with Charles D. Seeberger and jointly won first prize at the Paris Exhibition in 1900. Seeberger's role had been to buy George A. Wheeler's earlier moving walkway patent, and also to have merged the

Variable speed controls save energy by reducing the nominal speed of an escalator to around 0.2 m/sec when it is not being used.

words 'scala' – meaning 'step' in Latin – with Otis' own 'elevator'. An escalator after all, is essentially just an angled conveyor belt pulling a series of steps in a constant loop. At the top and bottom, these steps are aligned to form level landings for ease of access and dismount. But there have been constant improvements to the technology over the years and the Holy Grail of commercially viable, variable speed escalators seems just around the corner. (Actually, spiral escalators – escalators going round corners – are installed in Caesar's Palace, Las Vegas not to mention the fabled 'Reno double spiral continuous moving track' built at Holloway Road tube station in 1906).

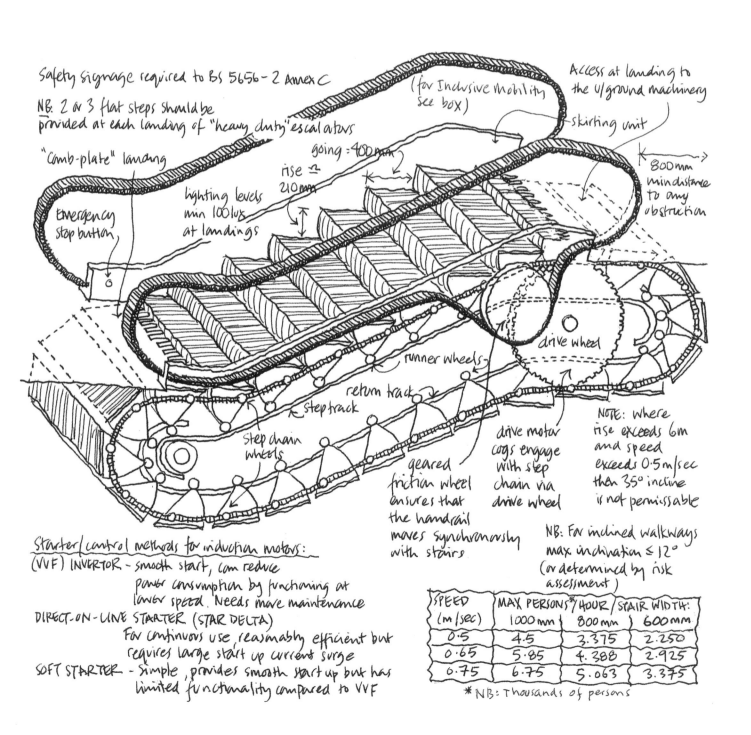

Safety signage required to BS 5656-2 Annex C

NB: 2 or 3 flat steps should be provided at each landing of "heavy duty" escalators

"Comb-plate" landing

Emergency stop button

going : 400mm

rise 210mm

lighting levels min 100lux at landings

(for Inclusive mobility see box)

Access at landing to the U/ground machinery

skirting unit

800mm min distance to any obstruction

drive wheel

runner wheels

return track

step track

step chain wheels

geared friction wheel ensures that the handrail moves synchronously with stairs

drive motor cogs engage with step chain via drive wheel

NOTE: where rise exceeds 6m and speed exceeds 0.5m/sec then 35° incline is not permissable

NB: For inclined walkways max inclination ≤ 12° (or determined by risk assessment)

Starter/control methods for induction motors:
(VVF) INVERTOR – smooth start, can reduce power consumption by functioning at lower speed. Needs more maintenance

DIRECT-ON-LINE STARTER (STAR DELTA)
For continuous use reasonably efficient but requires large start up current surge

SOFT STARTER – Simple, provides smooth start up but has limited functionality compared to VVF

SPEED (m/sec)	MAX PERSONS*/HOUR/STAIR WIDTH:		
	1000mm	800mm	600mm
0.5	4.5	3.375	2.250
0.65	5.85	4.388	2.925
0.75	6.75	5.063	3.375

*NB: Thousands of persons

VARIABLE SPEED

In 2003, a new variable speed travelator – a *trottoir roulant rapide* (TRR) meaning 'fast rolling pavement', also known commercially as 'Gateway' – was trialled at Montparnasse Metro in Paris, and more recently, Pearson International Airport in Toronto has had one installed (although the Pearson one is a pallet type rather than the belt type). French manufacturer CNIM – which supplies the escalators to London Underground – has safely and successfully accelerated 'pedestrians' from the usual 0.75 m/sec travelator speed up to 2.25 m/sec over a 200 metre track. This is the equivalent of 8 km/hr (which is less than the initial trial speeds of 10 km/hr because there were too many accidents), but is equivalent to more than twice normal walking speed. It is also half the average speed of a double decker in London. If these TRRs were installed across the city at pavement level, running for the bus could be a thing of the past.

Stepping onto a fast-moving travelator is fairly hazardous and so the Montparnasse high speed system uses small rollers instead of the traditional comb plate (the top and bottom landings that demarcate the start of the ride). Stepping onto the rollers, passengers are pulled along by holding the handrails. You are advised not to walk, so that your feet glide over a transition plate and onto the speeding travelator thus accelerating from the rollers' 0.75 m/sec into 2.25 m/sec. CNIM believes that for distances of around 500 m, the system will compete with the shuttle network.

GENERAL RECOMMENDATIONS FOR INCLUSIVE MOBILITY ON ESCALATORS*

- Maximum speed: 0.75 m per second (but 0.5 m per second is preferable especially where passenger use is relatively light)

- Preferred angle: 30 degrees

- Minimum width: 800 mm

- Maximum width: 1100 mm (also for inclined travelators >6 degrees and <12 degrees)

- Maximum riser: 240 mm (although 210 mm is preferable if the escalator is for designated use as an emergency exit when stationary)

- Treads: matt, non-reflective finish with 55 mm contrasting colour band as tread edge demarcation

- Handrail: 900 –1100 mm above step nosing extended a minimum of 300 mm beyond the ends of the escalator – it should be colour contrasted and be synchronised to move with the escalator

- Landings: a level space on the approach to the escalator at the top and bottom, at least 2 m and 1.6 m clear respectively (but can be as much as 10 m in heavily trafficked areas)

- Audio/visual aids: audible notification at the beginning and (just before) the end of the escalator and a minimum of 300 lux throughout (brighter at top and bottom, and giving good definition between treads and risers)

- Headroom: minimum 2300 mm vertical clear height above the pitch of the escalator

- Disabled: escalators are not appropriate for wheelchair users, etc and where there are substantial changes in level, a lift should be provided (clearly signposted as an alternative to the escalator)

- In emergencies: escalators and passenger conveyors should be provided with devices capable of being readily operated that, when activated, can bring the equipment to a controlled halt in such a way that passengers will be able to maintain their balance – the location of the emergency control devices and stopping performance, etc. of the equipment should meet the recommendations of BS EN 115: 1995

* Department for Transport (2002) *'Inclusive mobility. A guide to best practice on access to pedestrian and transport infrastructure',* DfT.

French trials of variable speed travelators have accelerated pedestrians to half the average speed of a London double-decker bus.

There's a long list of patents pending by those waiting to improve on Gateway's success. Mitsubishi are trying to vary the speed without relying on transition plates by using larger diameter pulleys at the escalator's edges which force the rollers supporting the steps to take wide turns, slowing them down. NKK has an ingenious travelator system that expands as it moves from normal to variable to high-speed zones.

However, 'variable speed' applied to escalators can have other meanings.

Energy saving: Variable speed controls save energy by reducing the nominal speed of an escalator to around 0.2 m/sec when it is not being used. This type of specification is now being added as standard to manufacturers' specifications. The Kowloon and Canton Railway system, for instance, has travelators that detect when passengers are not using it, slowing it by 30 per cent and reducing its electricity consumption by 56 per cent. Essen has a self-starting escalator activated by passengers walking past a sensor panel.

Torque control: Escalators have to be designed so that they don't slow down as soon as people step on them, i.e. when the load increases. In some manufacturing industries, conveyor belts have computer-operated direct torque control devices that can calculate the torque needed 40,000 times per second so that incredibly fast reaction times – and hence greater efficiency and reduced wear – can be maintained in the system. Escalators don't yet have that level of sophistication but torque control devices ensure that the power adjusts to cater for variable load.

PLEASE STAND ON THE . . .
The only remaining variable is the location for standing and walking, each altering the eccentric loading on the mechanism. Here in the UK, we stand on the right and walk on the left. The same is true of Washington, Boston, Hong Kong and Moscow escalators, but the opposite in Singapore, Australia and New Zealand. In Tokyo passengers stand on the left but on the right in Osaka.

For those of you who are confused, the American Elevator Escalator Safety Foundation has been set up to educate the public on the safe and proper use of escalators. Advice number one: 'Step on and off carefully'. For further information, visit: www.eesf.org

References

BS 7801 (2004) *'Escalators and moving walks – Code of practice for safe working on escalators and moving walks'*, BSI.

BS EN 115 (1995) *'Safety rules for the construction and installation of escalators and passenger conveyors'* (AMD 10030) (AMD 15436), BSI.

RECOMMENDED READINGS
BS 5656-2 (2004) *'Escalator and moving walks – Safety rules for the construction and installation of escalators and moving walks – Part 2: Code of practice for the selection, installation and location of new escalators and moving walks'*, BSI.

Chartered Institution of Building Services Engineers (2000) *'Transportation systems in buildings, 3rd edition, CIBSE Guide D'*, CIBSE.

Lift and Escalator Industry Association: http://www.leia.co.uk/

The Lift and Escalator Industry Association (2003) *'Personal lifting equipment sales, installation and aftercare – Code of Practice'*, LEIA.

25: Fire Detection
This building is alarmed

According to the 'Fire Statistics, United Kingdom', in 2002, there were nearly 280,000 false alarms from fire detection systems. In 2007, there was a false fire alarm on board the £50 billion International Space Station caused by a 'software problem'. Rumours that they hadn't replaced the battery were unfounded. But back on the ground, how should we specify fire alarm systems?

Here's how the unintended consequences of risk assessments work. Under the Regulatory Reform (Fire Safety) Order 2005 (RRO), a responsible person must carry out a risk assessment on all premises and buildings other than dwellinghouses in single occupation (see Shortcuts: Book 2), the purpose of which is to ascertain the hazards and risks present. In turn, the Fire and Rescue Service may visit the property, but they will not carry out a risk assessment; they will simply audit the procedures in place. Fair enough, you might think. Unfortunately, as a consequence, it seems that landlords and letting agents are insisting that fire extinguishers be removed from existing communal flats and not be installed in new ones, because they pose a fire risk. Why? Well, one risk assessor on a residential scheme in Dorset is quoted as saying that an extinguisher could cause a hazard if the person using it has not been trained; the risk assessment showed that residents should be encouraged to run away rather than stop to fight a fire. In this scenario, operating a fire extinguisher is deemed to be an uninsurable health risk.

There should be at least one smoke alarm on every storey of a dwellinghouse and a heat detector should be provided in the kitchen if there is no physical separation between it and the staircase.

The immediate effect of the Regulatory Reform (Fire Safety) Order 2005 has been an increase in the specification of sprinklers and automatic fire detection systems.

If that logic isn't bad enough, fire chiefs in Manchester are telling businesses that they must assess the validity of any activated fire alarm to confirm that it is not a false alarm. In other words, a designated responsible person must witness signs of a genuine fire before the fire authority will send out a fire tender. Given that false alarms distract the fire service from attending real fires, this might appear to be a sensible policy, but whether this will alleviate or exacerbate fire casualty figures, only time will tell. The immediate effect of the RRO has been an increase in the specification of sprinklers and automatic fire detection systems.

BS 5839-6: 2004 'Fire detection and fire alarm systems for buildings. Code of practice for the design and installation of fire detection and alarm systems in dwellings' recommends that the very least the designer should do is to specify the grade of fire alarm system. This should also be stated on the installation and commissioning certificate. These grades, in brief, are as follows:

Grade A – Highest specification, which includes detectors, sounders, controls and indicating equipment to BS EN 54-2, and a power supply to BS EN 54-4

Grade B – Fire alarms and detectors (other than smoke and heat alarms), controls, indicating equipment and power supply

Grade C – Fire detectors and sounders (these may be combined in a smoke alarm), a common power supply (i.e. mains and standby) and central controls

Grade D – One or more mains-powered smoke alarms (perhaps with additional mains-powered heat alarms) with integral standby supply

Grade E – One or more mains-powered smoke alarms (perhaps with additional mains-powered heat alarms) the standby supply is not obligatory

Grade F – One or more battery-powered smoke alarms (perhaps with additional battery-powered heat alarms)

For domestic applications, it is recommended that a risk assessment be carried out rather than the designer assuming that the lowest specification grade is automatically acceptable. For example, in a small single-family domestic refurbishment, a basic Grade F system may seem to be suitable. However, if there is any doubt about the ability of the occupier to maintain battery-operated smoke alarms soon after a low-battery warning signal is given, then Grade E should be adopted. Where there is a likelihood that the mains electricity might be turned off because of an inability of the occupier to pay the bills, then Grade D should be specified. The object, after all, is to protect life and limb.

BS 5839-1 'Fire detection and fire alarm systems for buildings. Code of practice for system design, installation, commissioning and maintenance' introduces further category classifications which relate to the locations and the extent of the system. For domestic situations, etc, these are as follows:

Category LD1/Category PD1 Installed in all areas of the building (other than toilets, bathrooms, etc.) to offer the earliest possible warning of fire, and the maximum length of time for escape

Category LD2/Category PD2 Installed only in high-risk areas and escape routes

Category LD3 Installed only in escape routes

In these categories, the designation:

'L' represents systems for the protection of life

'P' represents systems for the protection of property

'D' indicates 'dwellings'.

Note: As far as BS 5839 is concerned, a dwelling is a single household where fewer than seven people reside. Therefore, a home comprising a family unit of two parents and their quintuplets no longer constitutes a dwelling, and the specification has to change accordingly to one with no prefix 'D' attached to the Category.

New-build dwellinghouses must have a minimum fire detection and fire alarm system to

Grade D Category LD3. Two-storey dwellinghouses, excluding basement storeys (with any storey of 200 m³ or more), should be fitted with a Grade B Category LD3 system. Three-(and more)-storey dwellinghouses, excluding basement storeys (with any storey of 200 m³ or more), should be fitted with a Grade A Category L2 system.

For the majority of homes, LD2 systems are deemed to provide appropriate coverage. In this category, detectors are located in the circulation areas, the living room and the kitchen. Document B 'Fire Safety' adds that there should be 'at least one smoke alarm on every storey of a dwellinghouse and a heat detector should be provided in the kitchen if there is no physical separation between it and the staircase. When specifying the Grade and Category of a system, the locations of detectors and alarms need to be clearly stated.' However, Simon Ham, author of 'Legislation Maze: Fire' notes that over-complicating construction drawings with detailed information on fire systems should be avoided where possible – unless the locations of fire detection devices form an intrinsic part of the design. He advises that a note specifying, for example, that 'an L1 fire alarm and detection system complying with BS 5839: Part 1 be installed' will normally suffice. Too much detail may compromise the approach of any design and installation contractor that implements the work.

The following categories relate to buildings other than dwellings:

Category L4 Installed within circulation spaces, corridors and stairways to enhance the safety of occupants by providing warning of smoke within escape routes

Category L5 Installed for fire detection in only part of a building (although it may be part of a more extensive fire protection system)

Category M Systems that are manual and therefore incorporate no automatic fire detectors

Smoke detectors These may be used in any room or part of a dwelling (notwithstanding kitchens, bathrooms and shower rooms) but should be avoided in locations that might give rise to a false alarm. Smoke detectors comprise either one or other or a combination of both of the following:

- **Ionisation chamber smoke detectors** These activate when particulates reduce the electrical current between two electrodes. Most sensitive to small particles found in rapidly burning, flaming fires. Empirical data indicates that this type is not appropriate for use in kitchens.

- **Optical smoke detectors** These detect the dispersion of light caused by smoke particulates. Most sensitive to the 'large' particles found in dense smoke.

NOTE: A risk assessment must be carried out to ascertain the most appropriate detection device
- LD1: some relaxation in the overall structural fire integrity of the building may be allowed
- PD1: detection devices in loft/basement may be omitted if electric permanently disabled *

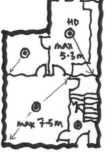  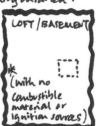

LD1/PD1 category

NOTE: Under PD category systems – any special areas containing high value goods, say, must be protected by fire-resisting structure including FR doors. Where this is not possible the system should be upgraded to PD1

LD2/PD2 category ⊚ SMOKE DETECTOR ⊚ᴴᴰ HEAT DETECTOR

NOTE: A smoke detector may not be the most appropriate device for kitchen fires

max 7.5m from smoke detector to any point

A risk assessment should be carried out to assess the need for additional HDs and SDs in bedrooms (carbon monoxide detectors may also be appropriate)

LD2 risk assessments should take account of response times as well as the need to minimise false alarms
PD2 risk assessments should locate detection in areas where ignition sources, or easily ignitable materials exist

LD3 category detection system

SMOKE DETECTORS: 25-600mm below soffit
HEAT DETECTORS: 150mm below soffit

If open plan kitchen/diner SD must be included to cover that area
Ditto: Stair/living room

⊚ SD
SMOKE DETECTOR

minimum distance from SD to bedroom door ≤ 3m

SD to have max 7.5m distance to any point in area of coverage

NOTE: wall mounted SD is permissible if hall is less than 10m long or wide and less than 50m²

Those installed in circulation areas such as hallways, staircase landings and corridors, should be of the optical type unless they are assessed to significantly increase the rate of false alarms than would otherwise be the case with ionisation chamber detectors.

Heat detectors Heat detectors comprise either one or other of the following (although these can be combined for maximum effectiveness):

- **Point detectors** These respond to the temperature of gases in the immediate vicinity of a single point location. These are the only type appropriate for use in dwellings, but should not be used in circulation areas, hallways, staircases, landings, etc.

- **Line detectors** These respond to the temperature of gases in the vicinity of a 'line'. Heat detectors are not sufficient, especially when positioned in a bedroom of potential fire origin, to save the life of a sleeping person. However, where heat detectors are recommended, a sprinkler head which activates the fire alarm system can be deemed to be a heat detector for the purposes of BS 5839: Part 6.

FIRE TERMS

Automatic fire alarm system automatically detects a fire, sets off an alarm and other actions (sprinklers, etc.). The system may include manual call points.

Conventional fire alarm describes a two-state fire detector – having a normal state and an alarm state. In conventional systems, devices are wired into a circuit or 'zone' of up to 20 detectors. An alarm is indicated by zone on the fire alarm panel.

Carbon monoxide (CO) fire detector is a device incorporating an electrochemical cell which senses carbon monoxide (CO) rather than levels of smoke or any other combustion products.

Fire alarm transmission link is a connection for transmitting fire signals and fault warnings from protected premises to a central (fire alarm) station or to a control room.

Fire detection system is a system of fixed apparatus, normally part of an automatic fire alarm system, in which fire detectors and a control panel (described in standards as 'Control and Indicating Equipment' [CIE]) are employed for automatically detecting fire and initiating other action as arranged.

Low fire risk area/room contains little or no combustible material (such as furniture, fittings, storage or linings) and no ignition sources, in which any foreseeable fire is unlikely to spread such as to present any significant threat to escape by occupants or damage to property.

Heat detector is a thermistor device (a resistor varying according to its temperature) to detect abnormally high temperature changes.

Mimic diagram is a representational map of the protected premises, such that an indication of the fire alarm system can be rapidly related to the layout of the premises.

Soak period is the period after a fire alarm system has been commissioned, but prior to handover, during which the system's performance in relation to false alarms and faults is monitored.

Zone is the subdivision of the protected premises, in which the fire alarm warning can be given separately, and independently, of a fire alarm warning in any other alarm zone.

References

British Approvals Service for Electronic Equipment in Flammable Atmospheres (BASEEFA) is the approval body, under the HSE, for products designed for use in hazardous areas. www.baseefa.com

British Fire Protection Systems Association Limited (2003) 'Mains Powered Fire Alarm Systems with No Standby Power Supply', Factfile 06, BFPSA.

BS 5839 'Fire Detection and Alarm Systems for Buildings':
    Part 1 (2002) 'BSI Code of Practice for System Design, Installation, Commissioning and Maintenance' (AMD 15447), BSI.
    Part 4 (1998) 'Fire detection and fire alarm systems. Power supply equipment' (AMD 14596) (AMD 16676), BSI.
    Part 5 (1988) 'Specification for Optical Beam Smoke Detectors', BSI.

BS EN 54 'Fire Detection and Fire Alarm Systems':
    Part 1 (1996) 'Fire Detection and Fire Alarm Systems. Introduction', BSI.
    Part 5 (2001) 'Heat Detectors. Point Detectors' (AMD 14338), BSI.
    Part 7 (2001) 'Smoke Detectors. Point Detectors using Scattered Light, Transmitted Light or Ionization' (AMD 14339) (AMD 16539), BSI.
    Part 10 (2002) 'Flame detectors. Point detectors' (AMD 16127) (AMD Corrigendum 16593), BSI.

Loss Prevention Certification Board (2006) 'Requirements and Testing Procedures for Radio Linked Fire Detection and Fire Alarm Equipment. Issue 1.0 Dated 23/01/2006', Loss Prevention Standard 1257, LPCB.

RECOMMENDED READINGS
BS 5446-2 (2003) 'Fire Detection and Fire Alarm Devices for Dwellings. Specification for Heat Alarms', BSI.

BS 5839 'Fire Detection and Fire Alarm Systems for Buildings':
    Part 1 (2002) 'Code of Practice for System Design, Installation, Commissioning and Maintenance (AMD 15447)', BSI.
    Part 6 (2004) 'Code of Practice for the Design, Installation and Maintenance of Fire Detection and Fire Alarm Systems in Dwellings' (AMD 15447), BSI.

BS EN 54 'Fire Detection and Alarm Systems for Buildings':
    Part 3 (2001) 'Fire Alarm Devices. Sounders' (AMD 14341) (AMD 16462), BSI.
    Part 4 (1998) 'Power Supply Equipment' (AMD 14596) (AMD 16676), BSI.

Ham, S. (2007) 'Legislation Maze: Fire', RIBA Publishing.

The Fire Industry Association (FIA) is a new trade association formed by the merger of the Fire Extinguishing Trades Association (FETA) and British Fire Protection Systems Association (BFPSA). www.fia.uk.com

26: Light Emitting Diodes
The market for LED balloons

If you believe the hype, the light emitting diode (LED) is variously going to save money, save the electrical industry and save the planet. Currently accounting for only a tiny percentage of the global lighting market, their value is expected to grow to $1Bn by 2011. Even though that'll still be just one per cent of total lighting sales, LEDs are becoming increasingly influential.

A light emitting diode (LED) comprises an electrical circuit encased in a plastic, epoxy resin or ceramic housing. The diode is a two electrode component that ensures current flows in one direction only.

The complete LED consists of a layer of electron-rich material separated from a layer of electron-deficient material on a semiconductor base. This is known as 'doped' material, i.e. that which has been bombarded with other particles to give it either a positive or negative characteristic. When current is passed through the diode, the atoms in one material are excited, releasing energy. During this energy release, light is created.

> Manufacturers still quote an LED lamp life of over 100,000 hours; however, as soon as that LED is encased in a fitting the lamp life reduces dramatically.

Colour temperature is a concept based on Max Planck's theory of the black body full radiator which means that the hotter it is the more blue it becomes.

BACKGROUND

The first commercially usable LEDs were developed in the 1960s by combining three primary elements: gallium, arsenic and phosphorus (GaAsP) to obtain a red light source. These were not very bright and were primarily used as indicators on clock faces, microwaves, etc. In the 1980s a new material, GaAlAs (gallium aluminium arsenide), was developed that made the LED ten times brighter, ensuring that LEDs could begin to be considered as a potential 'light source'.

By the early 1990s the first high-brightness blue LED was demonstrated but even now, it lacks the equivalent power of its red and green counterparts. Thus, a white light LED (which commonly uses the blue LED as the source light and transforms its UV output into white light via a phosphorous coating) – the holy grail of the residential/domestic sector – has not yet been perfected.

Here are a few other issues to be aware of when designing with LEDs.

1. Light output

A 'cool white' 1W LED will only produce around 50–60 lumens which means that, at present, LEDs have to be designed in clusters in order to get any decent lumen levels. Such a low lumen output is a factor of contemporary LED technology and some manufacturers are simply increasing the wattage to get more light output, thus negating the 'low energy' argument for using LEDs in the first place.

Advances in LED technology are raising expectations that the lighting industry is struggling to keep up with. For instance, a single-chip white LED with an output of 1050 lumens (equivalent to a standard household lamp) and an efficacy of 72 lumens per watt has been developed but it has yet to prove itself in site conditions.

2. Control

Unlike many other light sources, LEDs:

- are fully dimmable from 100 to 0 per cent

- are almost infinitely switchable

- give instantaneous 100 per cent output.

LEDs should usually be specified with a DMX (digital multiplexing) dimming protocol (a protocol is the way that intelligent systems and switches communicate with each other). This is commonly used in theatres, allowing control and scene setting of hundreds of LED fittings from one central location.

3. Heat

A poorly considered LED luminaire will get hot, and this can lead to output degradation, colour shift or even failure. An LED is not like a standard 'bulb' which screws into a fitting, but more like a chip that needs to be built into a housing. It tends to be the mass of this housing that acts as the heat sink.

The light output of an LED is dependent upon input current and whilst at 1 amp you can get upwards of 180 lumens from some LEDs, you would need a great deal of heat sinking to dissipate the internal heat from such a fitting.

4. Lamp life

Some manufacturers still quote an LED lamp life of over 100,000 hours, based on testing in conditions of regulated voltage, regulated temperature, etc. However, as soon as that LED is encased in a fitting, the lamp life reduces dramatically and light fitting manufacturers rarely test their products. Thus, when subject to uncontrolled external conditions, 20,000 to 40,000 hours of LED lamp life is more realistic. (This is still approximately two to five years continuous use.)

The lumen depreciation over time is significant even for high output white LEDs. The lumen output of older 5 mm white LEDs can reach 50 per cent of initial levels after only 6000 hours.

5. Colour – RGB

Recent developments in theatre design and DMX controls have created the possibilities for excellent colour mixing possibilities. By altering the red, green and blue (RGB) intensities, over 16.5 million colours can be achieved. Upon final commissioning this means an RGB LED system can be tweaked to give a colour that everyone in the design team is happy with – from strong, saturated colour to composite white light.

6. White light

It seems that the lighting industry will soon lose the option of using incandescent lamps and they are desperate to find white light replacements. Composite RGB white is OK for feature lighting but lacks the depth of colour and colour rendering of a proper white light solution. Even though a decent RGB white light is scheduled for the market in 2010, the efficiency of the linear fluorescent tube is going to be very hard to beat. Colour temperature is a concept based on Max Planck's theory of the black body full radiator which produces different colours of light as it is heated to different temperatures. Put basically, the hotter it is the more blue it becomes (like daylight) and the cooler it is the more yellow it becomes like a candle. This colour temperature is measured in Kelvin (Celsius plus 273 degrees; so that 21°C is the equivalent of 294 K). Daylight fluorescent lamps are 6000 K and warm white is 3000K. White LEDs are available in a range of colour temperatures (2760-10,000 K) – but fall short on colour rendering and lamp life.

7. Colour rendering

RGB white mixes colour from very narrow wavelengths of light and lacks the depth of colour to provide a suitable colour rendering for anything other than feature lighting. White LEDs are also lacking when it comes to colour rendering with a typical colour rendering index (CRI) of 60–70. (Note: CRI is a relative comparison between a lamp source and a reference source at the same temperature. A CRI value greater than 80 is suitable for indoor lighting; more than 90 is suitable for visual inspection tasks.)

To overcome this, some manufacturers are making luminaires that mix colour temperature LEDs in order to obtain a suitable colour rendering for the application. This isn't an ideal solution – the LED should have the appropriate colour rendering capabilities before it leaves the lab.

8. Photometric performance

Photometric data explains what the light pattern is, and roughly how it will behave. It's hard to believe, but most LED light fitting manufacturers do not even test how the light performs.

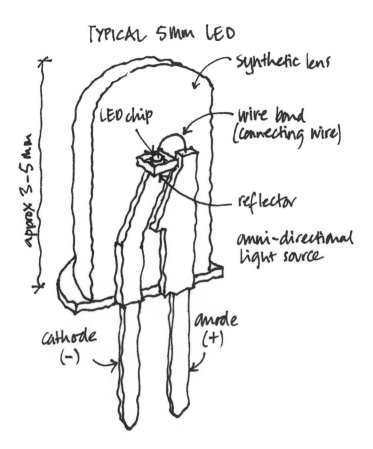

TYPICAL 5mm LED

synthetic lens

LED chip

wire bond (connecting wire)

approx 3-5 mm

reflector

omni-directional light source

cathode (-)

anode (+)

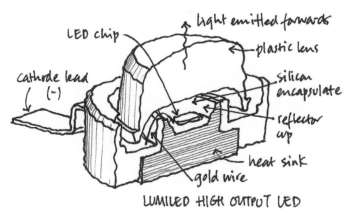

LED chip

light emitted forwards

plastic lens

cathode lead (-)

silicon encapsulate

reflector cup

heat sink

gold wire

LUMILED HIGH OUTPUT LED

LEDs and Gentlemen...

FURTHER INFORMATION IS AVAILABLE FROM:

Association of Lighting Designers, 01707 891 848, office@ald.org.uk

Commission Internationale de l'Eclairage, Austria, +43 1 714 31870, ciecb@ping.at

Hoare Lea Lighting, 020 7890 2500, www. hoarelealighting.com

Industry Committee for Emergency Lighting, 020 8673 5880, info@icel.co.uk

Institution of Lighting Engineers, 01788 576 492, info@ile.org.uk

International Association of Lighting Designers, USA, +1 312 527 3677, iald@iald. org

Lighting Association, 01952 290 906, enquiries@lightingassociation.com

Lighting Industry Federation, 020 8675 5432, info@lif.co.uk

Society of Light and Lighting, 020 8675 5211, sll@cibse.org

9. Energy efficiency

A standard 1W lumiled warm white LED gives approximately 25 lumens/watt. This would not meet Building Regulations Approved Document L's requirements, which range from a minimum of 40–50 lumens/watt for anything other than display lighting. Where LEDs have been a success in energy terms is in relation to feature lighting, which previously would have used high wattage units. Designers are learning that, in the external environment, they can be bold, use less light and still make a good impact, especially when colour is used.

10. The future

Despite the many drawbacks, the LED is currently the most promising future light source.

Recent developments are encouraging and, if these continue at the current pace, LEDs should be in a position to provide a 'green' alternative (i.e. environmentally friendly alternative) in the foreseeable future. For this to unfold, certain things need to happen:

- White LED needs to be a viable, cost-effective alternative to tungsten halogen and, to a lesser extent, to compact fluorescent. It will need to gain considerably greater light output, colour stability and colour rendering without resorting to higher wattages.

- Luminaire manufacturers need to incorporate the highest efficacy LEDs and suitable reflectors or 'secondary optics' within purpose-built LED luminaires. These must be designed to complement the light source, not just to fit a light source within an existing product.

The pace of LED innovation is not likely to slow, so the modern luminaire manufacturer will not be relying on static technology as they have had to with previous lamp types. Architects ought to keep a close eye on contemporary LED research and development and make more demands on manufacturers to supply appropriate useable data based on real site condition test results.

Many thanks to Hoare Lea Lighting for their assistance in this Shortcut.

References

BS EN 60598-1 (2004) *'Luminaires – Part 1: General requirements and tests'* (AMD 16925) (AMD Corrigendum 17471), BSI.

BS EN 61347-1 – A1 and A2 (2001) *'Equipment for lamps, Part 1: general and safety requirements'*, BSI.

BS EN 120001 (1993) *'Harmonized system of quality assessment for electronic components. Blank detail specification: Light emitting diodes, light emitting diode arrays, light emitting diode displays without internal logic and resistor'*, BSI.

Proceedings of the CIE Symposium 2004 on LED Light Sources (2004) *'Physical Measurement and Visual Photo Biological Assessment'*, 7–8 June 2004, Tokyo, Japan, CIE.

ZVEI *'Method for determining the life expectancies of LED modules in electric luminaries'*, German Electrical and Electronic Manufacturers Association, Frankfurt am Main.

RECOMMENDED READINGS

BS EN 61347-2 (2006) *'Lamp control gear. Part 2-13: Particular requirements for d.c. or a.c. supplied electronic control gear for LED modules'*, BSI.

BS EN 62384 (2006) *'DC or AC supplied electronic control gear for LED modules – Performance requirements'*, BSI.

BS IEC 60838-2 (2006) *'Miscellaneous lamp holders. Particular requirements. Connectors for LED-modules'*, BSI.

Building Applications Guide BG 3 (2005) *'Light-emitting diodes. A guide to the technology and its applications'*, BSRIA.

Dowling, K. (2005) *'Metrics for solid-state lighting'*, Leeds Magazine, May 2005.

Merck, D. (2004) *'Lighting and electrics'*, AJ Focus, May 2004.

27: How Lifts Work
Braking your fall

Amuse your friends and annoy your fellow passengers by putting your lift into 'Express' mode. On most modern lifts, by simply pressing the 'Door Close' and 'Floor' buttons at the same time, theoretically, your lift car will prioritise travel to the floor of your choice, bypassing the others.

After the Great Exhibition of 1851 in London, New York held its own Crystal Palace Exposition in 1853. It was here that Elisha Graves Otis demonstrated his new elevator safety device. Standing atop an exposed lift car, he was hoisted to a great height and then had an assistant cut the suspension rope with an axe. As the crowd gasped, the lift dropped only a few inches before being held by his invention – a 'toothed guide rail' – located on either side of the shaft. Three years later, he installed his first passenger lift in a New York City shop. One hundred and fifty years later, the company has estimated that it raises and lowers the equivalent of the population of the entire planet every nine days.

There are two distinct types of lifts (or elevators): electric traction and hydraulic. This Shortcut deals with the former (hydraulic lifts and car accessibility will be dealt with in later publications) and provides a general overview of how lifts function.

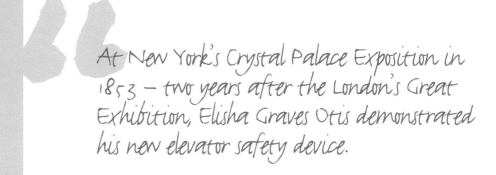

At New York's Crystal Palace Exposition in 1853 – two years after the London's Great Exhibition, Elisha Graves Otis demonstrated his new elevator safety device.

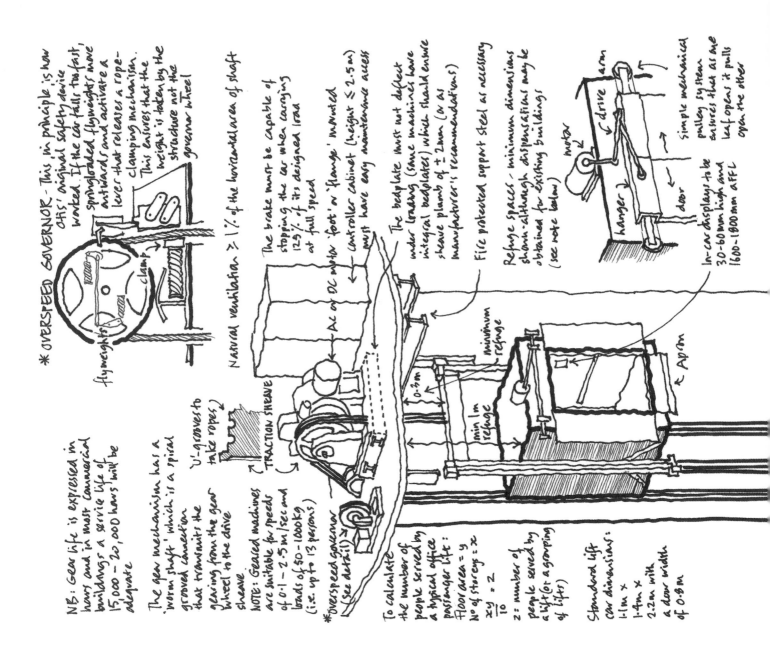

* OVERSPEED GOVERNOR - This, in principle, is how Otis' original safety device worked. If the car falls too fast, spring-loaded flyweights move outwards and activate a lever that releases a rope-clamping mechanism. This ensures that the weight is taken by the structure not the governor wheel

clamp

flyweights

Natural ventilation ≥ 1% of the horizontal area of shaft

NB: Gear life is expressed in hours and in most commercial buildings a service life of 15,000 – 20,000 hours will be adequate

The gear mechanism has a 'worm shaft' which is a spiral grooved connection that transmits the gearing from the gear wheel to the drive sheave.

NOTE: Geared machines are suitable for speeds of 0·1 – 2·5 m/sec and loads of 50 – 1000 kg (i.e. up to 13 persons)

'U'-grooves to take ropes

TRACTION SHEAVE

* Overspeed governor (see detail)

The brake must be capable of stopping the car when carrying 125% of its designed load at full speed

Controller cabinet (height ≤ 2·5 m) must have easy maintenance access

AC or DC motor 'foot' or 'flange' mounted

The bedplate must not deflect under loading (some machines have integral bedplates) which should ensure sheave plumb of ± 2 mm (or as manufacturer's recommendations)

Fire protected support steel as necessary

Refuge spaces - minimum dimensions shown, although dispensation may be obtained for existing buildings (see note below)

motor

drive

hanger

door

simple mechanical pulley system ensures that as one leaf opens it pulls open the other

In-car displays to be 30-60 mm high and 1600-1800 mm AFFL

minimum refuge

0·3 m

min 1m refuge

Apron

To calculate the number of people served by a typical office passenger lift:
Floor area = y
No of storeys = x
$$\frac{xy}{10} = z$$

z = number of people served by a lift (or a grouping of lifts)

Standard lift car dimensions:
1·1 m ×
1·4 m ×
2·2 m with a door width of 0·8 m

Within each of the electric traction lift types there are several variations, the main ones relating to direct (DC) or alternating current (AC), gearing mechanisms and motor speeds. Gearless machines are ones in which the motor rotates the sheave (or cable wheel) directly, whereas in geared machines the motor turns a gear shaft that rotates the sheave:

- Gearless machines comprise a motor (with a DC armature or an AC rotor), a drive sheave, a support framework to take the load(s) and a brake.

- Geared machines comprise traction sheave, gearbox, brake, motor and support framework.

- In both cases, sometimes a deflector sheave is used to alter the line of the cable drop.

Typically, all of the motorised equipment and the various control systems are housed in a dedicated machine room located above the elevator shaft (as distinct from the typical hydraulic lift layout) and often this machine room may protrude above the roof line. It is possible, with different pulley arrangements, for the machine room to be housed in locations other than the head of the shaft, but the consequent increase in rope length tends to give rise to self-load stretching and, hence, greater maintenance costs.

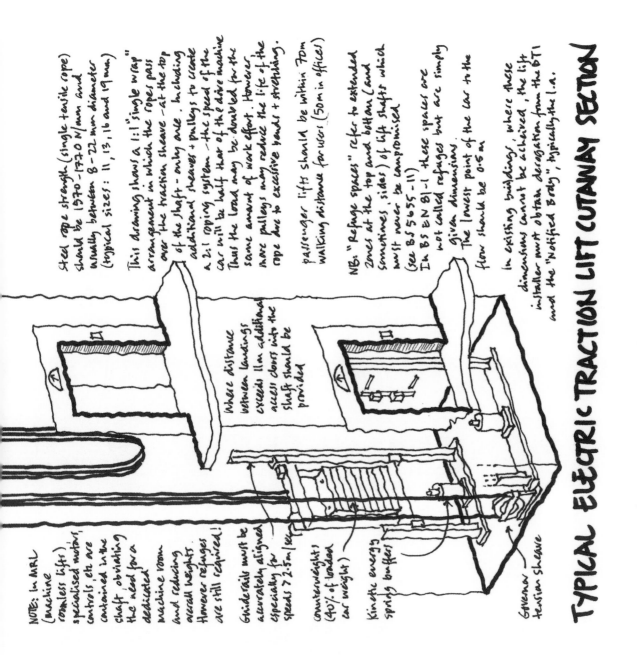

Steel rope strength (single tensile rope) should be 1570-1770 N/mm and usually between 8-22 mm diameter (typical sizes: 11, 13, 16 and 19 mm)

This drawing shows a 1:1 "single wrap" arrangement in which the ropes pass over the traction sheave - at the top of the shaft - only one. Including additional sheaves + pulleys to create a 2:1 roping system - the speed of the car will be half that of the drive machine. Thus the load may be derived for the same amount of work effort. However more pulleys may reduce the life of the rope due to excessive bends + stretching.

passenger lifts should be within 70m walking distance for uses (50m in offices)

NB: "refuge spaces" refer to extended zones at the top and bottom (and sometimes, sides) of lift shafts which must never be compromised (see BS 5655-11)
In BS EN 81-1 these spaces are not called refuges but are simply given dimensions.
The lowest point of the car to the floor should be 0.5 m

In existing buildings, where these dimensions cannot be achieved, the lift installer must obtain derogation from the BTI and the "Notified Body", typically the L.A.

NOTE: In MRL (machine roomless lifts) specialised motors, controls, etc. are contained in the shaft, obviating the need for a dedicated machine room and reducing overall heights. However refuges are still required!

Where distance between landings exceeds 11m additional access doors into the shaft should be provided

Guiderails must be accurately aligned especially for speeds > 2.5m/sec

counterweight (40% of loaded car weight)

Kinetic energy springs buffers

Governor tension sheave

TYPICAL ELECTRIC TRACTION LIFT CUTAWAY SECTION

Regardless of the location of the machine room, when regularly hoisting large loads, or for lifts reaching speeds greater than 4 m/sec, pulley arrangements are often the only option. Similarly, some manufacturers now supply machine-room-less (MRL) lift systems where the drive sheaves and pulleys, etc. are located within the shaft.

The Consortium Local Authorities Wales' 'Lift Services Specification' provides very useful guidance on the subject. However, non-specialist designers should be aware of various liabilities associated with the installation of lift technology. For example, even though lighting may be required in a lift shaft, The Lifts Regulations 1997 states that if the design and construction of a lift shaft contains 'any piping or wiring or fittings' that are not 'necessary for the operation and safety of that lift', then the 'person responsible for work on the building or construction' is liable to a hefty fine and three months imprisonment. The identity of that 'person responsible' is not completely clear.

In his book 'Faster: The Acceleration of Just About Everything', James Gleick notes that, in studies it was found that office workers regularly complained about waiting for 'ten minutes' for the lift to arrive when, in fact, timings showed that the lift had arrived within two minutes. He says that our expectation of speed frequently leads to a frustration with waiting – the original elevator traveled at 20 cm/sec but the lift in the world's tallest

The lift 'types' mentioned in the 'Metric Handbook' do not correlate to the official lift designations. In fact, lifts are divided into the following classifications:

Class I – designed for the transport of persons

Class II – designed mainly for the transport of persons but in which goods may be carried (BS ISO 4190 states that these lifts differ from Class I, III and VI lifts essentially by the inner fittings of the car, although this is not further explained)

Class III – designed for health care purposes, including hospitals and nursing homes

Class IV – lifts designed mainly for the transport of goods (freight) that are generally accompanied by persons

Class V – service lifts (US: dumbwaiters)

Class VI – specially designed to suit buildings with intensive traffic, i.e. lifts with speeds of 2.5 m/sec and above

Lift cars tend to cite passenger numbers, which, regardless of the actual weight of the population, translates as:

320 kg – 4 person

400 kg – 5 person

450 kg – 6 person

630 kg – 8 person

800 kg – 10 person

1000 kg – 13 person

building, Taipei 101 Tower, now holds the world record of 17 m/sec (around 38 mph). It contains a pressure control system to prevent the passengers' ears popping, a hi-tech balance mechanism to reduce vibrations to a minimum, and specially streamlined cars to reduce the noise of cars literally whistling past each other.

In the more sedate UK, BS 5655-6: 2002 'Lifts and service lifts' recommends that, in order to reduce 'buffeting' for lifts with a speed exceeding 2.50 m/sec, pressure release vents might be needed. A vent, equivalent to 1 per cent of the shaft area, must be provided at the top of lift shafts, but 'a single lift can require a vent of a minimum area of 0.3 m² with an additional vent area of 0.1 m² for each additional lift sharing a common well'.

Often, it is the speed of the car door opening system that is the biggest factor in the waiting time of the system. The car door motor located above the car is a relatively simple mechanism for converting a rotary motion into a linear motion to push the leaves open. Single-speed AC door operators are best suited for door openings up to 800 mm wide for low-density traffic. Medium-speed (two-speed) door operators for 1000 mm centre-opening doors can open in 3.3 seconds and close in 3.7.

With grateful thanks to CIBSE's *Transportation systems in buildings'.*

> *Office workers regularly complained about waiting for ten minutes for the lift to arrive when, in fact, timings showed that the lift had arrived within two.*

References

BS 8486: Part 1 (2007) *'Examination and test of new lifts before putting into service – Specification for means of determining compliance with BS EN 81. Electric lifts',* BSI.

BS 8486: Part 2 (2007) *'Examination and test of new lifts before putting into service – Specification for means of determining compliance with BS EN 81. Hydraulic lifts',* BSI.

Lienhard, J.H. (2003) *'The Engines of Our Ingenuity: An Engineer Looks at Technology and Culture',* Oxford University Press.

PAS 32-1 (1999) *'Specification for the examination and test of new lifts before putting into service. Electric traction lifts',* BSI.

PAS 32-1 (1999) *'Specification for the examination and test of new lifts before putting into service. Hydraulic lifts',* BSI.

RECOMMENDED READINGS
BS 5655: Part 6 (2002) *'Lifts and service lifts – Code of practice for the selection and installation of new lifts',* BSI.

BS 5655: Part 11 (2005) *'Lifts and service lifts – Code of practice for the undertaking of modifications to existing electric lifts',* BSI.

BS 5655: Part 12 (1989) *'Recommendations for the installation of new, and the modernization of hydraulic lifts in existing buildings',* BSI.

BS EN 81 Part 2 (1998) *'Safety rules for the construction and installation of lifts. Hydraulic lifts (AMD Corrigendum 10810) (AMD 15795)',* BSI.

BS ISO 4190-1 (1999) *'Lift (US: Elevator) installation – Part 1: Class I, II, III and VI lifts',* BSI.

BS ISO 4190-2 (2001) *'Lift (US: Elevator) installation – Part 2: Class IV lifts',* BSI.

Chartered Institution of Building Services Engineers (2005) *'Transportation systems in buildings',* 3rd edn, CIBSE.

Engineering Project Group and Marald Engineering Consultants (2007) *'Lift Services Specification',* Consortium Local Authorities Wales.

Statutory Instrument (1997) *'The Lifts Regulations 1997',* No. 831, TSO.

28: Toilets for Disabled Access
A general assessment of the needs for disabled access

How near to universally acceptable disabled access sanitary accommodation can designers get? And what are the real needs that have to be catered for? This Shortcut looks at some of the issues that need to be considered to satisfy the regulations as well as the (sometimes conflicting) needs of real building users.

The Disability Discrimination Act insists that designers provide disabled access facilities to the extent that it is reasonable so to do. Under the heading of what is 'unreasonable', financial constraints can be factored in as a variable. But even with this potential get-out clause, over the past 15 years or so, toilets designed for generic disability have become normal practice in most public buildings. Disability access has ceased to be a chore for architects and has almost become a mainstream consideration. Moreover, regardless of the access auditor, the disability mentor, the impairment officer or the risk assessor, building designers frequently attempt to provide better facilities than the basics outlined in Approved Document M: 'Access to and Use of Buildings' (AD M) or Scottish Technical Handbook 4: 'Access Within Buildings' and equivalents.

statistics showing that there are '5000 guide dog owners' is not so eye-catching as '2.5 million people with visual impairments'.

BS 6465-1:2006 gives user numbers (rather than densities) for a range of building types but does not refer to toilets for the disabled. Disabled toilet requirements are very clearly detailed in BS 8300.

However, with official statistics suggesting that 11.7 million people in the UK population are disabled, you can almost guarantee that you will never get an accessible toilet exactly right for everyone. Surely the faintly fatuous notion that around 20 per cent of the adult ought to be covered by the DDA doesn't help the discussion about the types of facilities needed by the severely disadvantaged sections of the public. The much-quoted figure of 1.5 million people with 'learning disabilities' for example, incorporates 40,000 people with *profound* learning difficulties, and consequently the real needs of this subset often tend to get lost in the generalised responses to the much higher global figure. The concept of learning difficulties, defined as not learning things 'as quickly as others', surely ought not to embrace equally those who have 'problems with co-ordination' or those with 'confusions between left and right' with those with severe mental illness.

Prioritising the generic figures can sometimes downplay severe disability and overcomplicate a more rational approach to disability access. After all, statistics showing that there are '5000 guide-dog owners' is not so eye-catching as '2.5 million people with visual or hearing impairments'. Actually, separating short-sightedness and total blindness might allow designers to focus attention on the specific levels of the problem. Then, we might ask, for example, how is a registered blind person expected to take a guide dog into the paltry sanitary facilities that still exist in shopping centres up and down the country?

As Selwyn Goldsmith, author of the seminal 'Universal Design' and 'Designing for the Disabled' has said, designers still have a wheelchair user *sans assistant* in mind when designing sanitary accommodation. In fact, many wheelchair users have carers and assistants and when you factor them in, or consider the less socially acceptable types of disability – mental as well as physical – then many WC cubicle designs still fall far short of the mark. Obviously the obligatory 1500 mm turning circle is reasonable allowance although extending this to 1700 mm where possible would better facilitate a carer. But, other issues could be improved upon. Here are a few:

- floor surfaces need to be non-slip (see Shortcuts: Book 2) but not overly reflective

- lobbies should be avoided if possible (notwithstanding the hygiene needs for a lobby onto food prep or storage areas)

- door closers should be easy to push against with a maximum 20 N force (consider using rising butt hinges)

- outward-opening doors with dedicated ironmongery

- the colour contrasts within the toilet compartment need to aid clarity but not be garish

- there should be level access

- shelving and equipment with rounded edges

- consistent lighting which must not be on a timer

An essential requirement is that grabrails must be fixed securely, with the substrate and fixings sufficiently rigid and sturdy to withstand a person's full body weight. Coat hooks can be fixed at various heights provided that the lowest one does not drape garments along the floor. Radiators, disposal bins and vending machines should, where possible, be recessed, and plumbing should not protrude into the space even if boxed in, in case it restricts manoeuvrability. Ensure that the WC seat remains upright when positioned against the backrest; that all equipment is 'heavy duty' and fixed accordingly (as opposed to typical domestic nut and bolt arrangements); and that dispensers, etc. are easily operable and within reach. Tap water temperatures should not exceed 41°C, be operated by lever grip and be of sufficiently low pressure to prevent splashing the floor. And my own personal gripe: spare toilet roll holders should not be located at ground level alongside the WC where the splashback effect is experienced the most.

The AD M, together with guidance by many local authorities, recommends that baby-change facilities should not be provided in accessible toilets or in changing rooms since this will restrict the facilities for disabled people. Changing and feeding babies often takes a long time and could cause a considerable wait for people with disabilities, So baby-change facilities should be catered for eleswhere where possible. However, unless

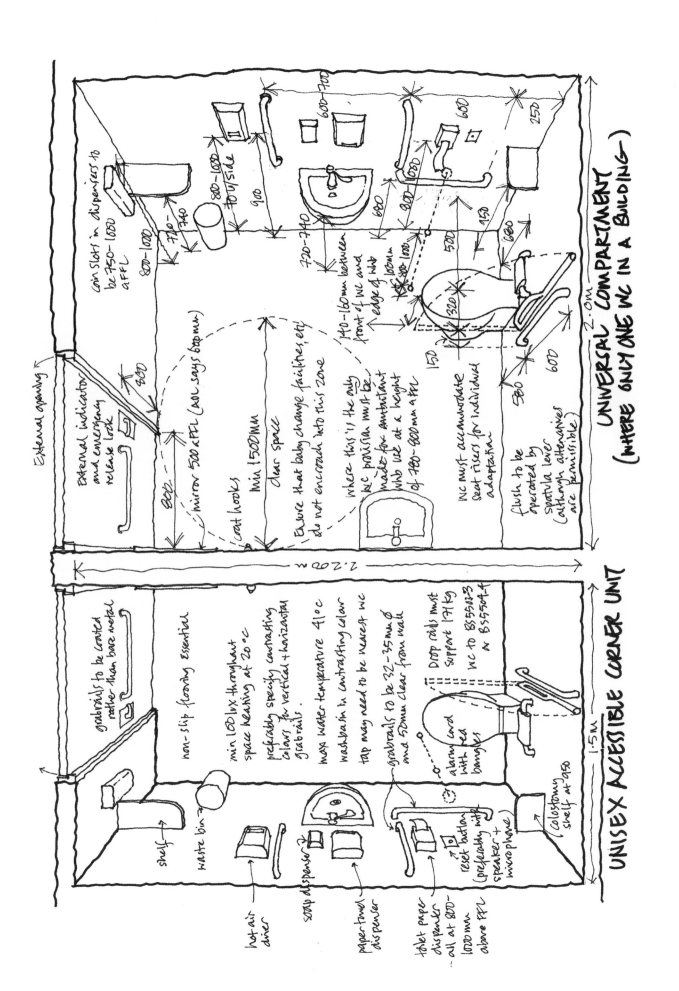

audited otherwise, the likelihood of disabled and baby change access occurring at the same time is slight – and so it is reasonable to optimise the compartment space. If located in disabled access toilets, baby-changing equipment should be flush-mounted so as not to impede the general circulation and turning circles within the compartment.

While the AD M specifies that the height to the top of washbasins should be 720–740 mm, care should be taken to specify shallow bowls to avoid obstruction to wheelchair users. Take note also of a conflict between this and the 780–800 mm height for washbasins for ambulant disabled (often those with chronic back pain). You can either specify to the letter of the legislation (see drawing) or a compromise can be sought: a) agree the optimum height with individual Building Control Bodies; b) provide an additional washbasin at the alternative height (as required by AD M, if this is the only WC in the building), or; c) recognise that not everyone is going to be totally satisfied and target your specification towards the more prevalent, or more debilitating disability.

But it's not just about access inside the toilet compartment; it's also about getting to the compartment in the first place, and travel distances from any one point should be 40 m max and on level surfaces where possible. The British Toilet Association recommends that the number of female cubicles should be equal to 2 × (male urinals + male cubicles). Goldsmith calculates on the basis of floor areas and recommends that the ratio be 3:1 (area of female compartments : area of male compartment) or even 4:1 in the case of theatres. BS 6465-1: 2006 'Sanitary installations' links overall toilet provision to the population densities (those figures that are used for fire regulations) although where the actual occupancy figures are not known, assume an office density of one person per 6 m². Otherwise, a standard allowance of 1 person per 10–14 m² would be acceptable for calculating office toilet provision. Please note though that the tables in BS 6465-1: 2006 give user numbers (rather than densities) for a range of building types but do not refer to toilets for the disabled. Disabled toilet requirements are very clearly detailed in BS 8300.

The unisex toilet shown in this Shortcut is designed for angled lateral transfer from one direction only (alternatives shown). Peninsular layouts allow transfer from either side but should only be specified in instances where skilled assistance is available. As the Centre for Accessible Environments says, there is a 'national reticence' about these issues, but bearing in mind that 'disabled men will mostly use WCs in public buildings as a urinal' ... approached head on, care should be taken to avoid obstructing legs and footplates on either side of the toilet.

The final point to note is that disabled facilities *can* be well designed. They do not need to replicate floor-to-ceiling tiled, sterile hospital toilets. Although not everyone can be catered for, consideration in the design and specification stages for likely user groups' difficulties – rather than ticking the boxes – will provide an environment beneficial to most users.

References

BS 6465-1 (2006) *'Sanitary installations – Part 1: Code of practice for the design of sanitary facilities and scales of provision of sanitary and associated appliances'*, BSE.

BS 8300 (2001) *'Design of buildings and their approaches to meet the needs of disabled people. Code of practice'*, BSE.

Department for Communities and Local Government (2004) *'Approved Document Part M: Access to and use of Buildings'*, NBS.

Goldsmith, S. (2000) *'Universal Design'*, Architectural Press.

Goldsmith, S. (1997) *'Designing for the Disabled: The New Paradigm'*, Architectural Press.

The Stationery Office *'Disability Discrimination Act 1995'*, TSO.

The Stationery Office *'Health and Safety at Work Act 1974'*, TSO.

RECOMMENDED READINGS

Centre for Accessible Environments (2004) *'Designing for Accessibility'*, RIBA Enterprises.

Centre for Accessible Environments (2004) *'The Good Loo Guide'*, RIBA Enterprises.

Department for Communities and Local Government (2008) *'Improving public access to better quality toilets – a strategic guide'*, DCLG.

29: Put That Light Out
Fire risk and portable firefighting equipment

In a fire, often it isn't the actual flames that do the damage. Frequently, it is the smoke contamination or the unavoidable water ruination caused by the fire service's intervention after the fire has taken a hold. Early and effective prevention of fire spread can minimise the damage resulting from a local blaze.

The front cover of the first edition of George Monbiot's book 'Heat' shows a representation of a fire extinguisher labelled 'How to stop the planet burning'. The implication is that simple, personal mitigation strategies are an essential step to preventing bigger, more dangerous conflagrations, and it is an apt illustration of the benefits of firefighting equipment. Extending the analogy to a global scale is debatable, but in relation to the trusty fire extinguisher, it is true that preventative action to combat domestic and non-domestic fires has long been recognised to be a sensible statutory necessity. However, on 1 October 2006, the statutory framework changed with the coming into force of the Regulatory Reform (Fire Safety) Order 2005 (RRO), doing away with 70 separate pieces of fire legislation and replacing them with a single structure. The RRO affects all buildings except domestic accommodation in single occupancy (although it does relate to communal areas in residential buildings).

Under the new regime, a fire certificate is worthless and the conditions of license no longer apply. Instead, fire safety becomes the province primarily of the 'responsible person'. In a workplace, this tends to be the employer or designated employee(s); in all other premises it refers to those with effective control of any part of the premises, such as the occupier and owner. Where the responsibility falls to more than one person, all reasonable steps have to be taken for each to work with the other(s).

The requirements of the RRO must be discharged via the ubiquitous risk assessment, identifying areas that need attention and those residual risks that need to be managed. Businesses employing more than five persons are under a duty to record these risk assessments, which may then be subject to monitoring by the fire service. Given that the fire service is stretched, it has confirmed that it will focus its attention on buildings that it adjudges to carry the highest risk (see tables). When carrying out an inspection, the fire officer will assess the quality of the documentation and the professionalism with which it is presented, in order to work out whether the visit ought to be long, detailed and hostile, or not.

The fire service no longer provides free advice – although that doesn't mean that they won't be helpful – but in general, detailed advice tailored to a specific property or business is now chargeable. The Association of British Insurers is worried that this 'could increase compliance costs for small businesses and their susceptibility to fire'. Indeed, while personal safety awareness seems to be improving, exemplified by the fact that fire deaths (less than 30 per annum) in non-dwellings are around half those of 15 years ago, the financial costs of a serious fire can be high. Total commercial losses amount to £2.5 million every day in the UK, and across all sectors the figure is adjudged by the Home Office to be as high as £7 billion, one-third of which is alleged to have been caused by arson.

ASSESSMENT AND MITIGATION

The duties of the responsible person comprise: making an assessment of the risks in order to prevent fires; taking practical measures to avoid or reduce the likelihood of a fire; and implementing transparent procedures that will enable the decisions to be audit-trailed.

Of the 33,400 fires in non-domestic buildings in 2004, over 80 per cent were confined to the room in which they started, highlighting the importance of good local containment strategies (such as portable fire extinguishers, sprinklers, fire doors, etc.) as well as the effectiveness of alarm systems and evacuation policies (see Shortcuts: Book 2). Even though the Fire Extinguishing Trades Association (FETA) and the Independent Fire Engineering & Distributors Association (IFEDA) have identified generic risk environments (see tables), individual risk assessments are required to identify specific hazards.

Once the risks have been highlighted, and residual risks (those that cannot be reasonably designed out or rectified) documented, a fire management strategy needs to be compiled. A key aspect of this normally includes the provision of portable fire extinguishing equipment, the use of which is reputed to save the UK economy over £500 million and to preserve around 25 lives per annum. As such, they are often an insurance requirement although there are some experts that argue that an evacuation strategy for a public building might be jeopardised by persons attempting to tackle a blaze by hand.

A building's fire management strategy requires that extinguishers be clearly visible, accessible and regularly maintained. If the floor plate in a multi-storey building is repeated, extinguishers should be placed in similar locations on each floor. For Class A fire risks (see specification criteria below), these should preferably be at designated fire points, which are areas containing, *inter alia*, fixed firefighting equipment (hose reels, etc.), directional signage, alarm calls and escape stairs. For other fire classes, extinguishers should be placed near the hazard, although it is essential that they are still positioned along the escape route.

Emergency lighting must be positioned such that it illuminates firefighting equipment and alarm call points, possibly in addition to that necessary to illuminate the escape route. If the extinguishers are partially hidden from view then the Health and Safety (Safety Signs and Signals) Regulations 1996 require that their location be indicated by (directional) signage.

SPECIFICATION CRITERIA

Currently, there is no single fire extinguishing agent that can be used on all types of fire and in order to work out which extinguisher is appropriate for the job, BS EN 2 classifies fires in accordance with their combustible materials, as follows:

- Class A: solid (usually organic) materials (wood, paper, cloth, plastics, etc.)

Deal with using water, foam or multi-purpose powder extinguishers (water and foam being the most common).

- Class B: miscible and non-miscible liquids (or liquefiable solids). Note: 'miscible' means that which can be mixed (with water).

FIRE EXTINGUISHERS FOR DIFFERENT FIRE CLASSES | TYPICAL EXTINGUISHER REQUIREMENTS

	A	B	C	D	⚡	F	communal accommodation	laboratories	large kitchen	small kitchen	plant room
likelihood of class of fire occurring (%)	31	23.3	1.5	n/a	43.6	0.6					
water (with or without additives)	X						9L		6L		
foam	X	X						6kg			6L (MPF)
dry powder	X	X	X		X						
CO₂	?	X	X		?		2kg	2kg	2kg		2-5kg
wet chemical (specialist)	X					X			6L for ⚡		
dry powder (specialist)				X							
hose reel	X						?				
fire blanket	X	X					?	1.1m²	1.1m²	1.1m²	

Note: MPF = multi-purpose foam

NOTE: ? = occasional use, check risk assessment and manufacturer's details

Tackle with sprayed foam (including multi-purpose aqueous film-forming foam (AFFF) carbon dioxide or dry powder extinguishers.

- Class C: gases

Use dry powder extinguishers, although care must be taken to halt the leakage of gas and/or any build-up of unburned gases.

- Class D: metals and powdered metals

These usually require trained personnel using special equipment to tackle them. Dry powder extinguishers may be appropriate subject to a dedicated risk assessment.

- Class F (this is a BS 7937 classification rather than BS EN 2): high temperature cooking oils or fats (of a higher risk than Class B)

A small number of specialist extinguishers are available for Class F fires involving cooking fats and oils (chip pans and deep fat friers, for instance), but these should only be used by specially trained personnel. Heavy-duty (as opposed to light-duty) fire blankets can also be appropriate. Cooking appliances cause 25 per cent of all fires in the workplace.

Note: Electrical fires are not given a fire classification (although occasionally referred to as Class E). Carbon dioxide extinguishers can sometimes be suitable for these fires, although the power must be isolated first.

Deferring to specific risk assessment documents, in general one extinguisher should be provided per 200 m² of floor space for Class A fire risks, with a minimum of one per floor. However, if the floor is around 90–100 m², just one 13A rated extinguisher may be used, provided that there is a maximum travel distance of 30 m to reach it. (Note: The designation 13A relates to the ability of the extinguisher to put out a Class A (wood) fire of given height and 1.3 m length. An 8B rating relates to an extinguisher's effectiveness in putting out 8 litres of Class B material.)

The use of a hose reel of sufficient reach – provided that only Class A fires are anticipated – may eliminate the need for water-type extinguishers. In certain locations such as laboratories, sand buckets may be useful to douse flammable liquids if they are used in conjunction with the recommended extinguisher.

> *Total commercial losses amount to £2.5 million every day in the UK, and across all sectors the figure is adjudged by the Home Office to be as high as £7 billion per annum, one-third of which is alleged to have been caused by arson.*

> *Once the risks have been highlighted and residual risks (those that cannot be reasonably designed out or rectified) documented, a fire management strategy needs to be compiled.*

COLOUR CODES AND SIZES

Extinguishers manufactured to the current BS EN 3-7: 2004 are predominantly red with a band of colour across the top designating the contents: Red (water), Cream (foam), Blue (powder), Black (CO_2), Yellow (wet chemical for specialist use). Properly maintained, extinguishers can have a life of some 20 years, and so the older versions may co-exist with the new versions in existing buildings. Where possible, this should be avoided. Also, halon extinguishers have been phased out in accordance with the 1987 'Montreal Protocol on Substances that Deplete the Ozone Layer'. Note: These extinguishers (usually green in colour when used in domestic situations) have been illegal since 2003 and need to be disposed of in officially designated sites.

Portable fire extinguishers must not weigh more than 20 kg as defined in the BS EN 3-7. However, HSE guidance on safe lifting practices implies that 20 kg extinguishers should not be operated by women (for whom a maximum lifting weight of 16 kg is recommended). Men are advised only to lift 20 kg if done with bent arms and from a height of the grip of approximately 250 mm. In general then, mounted extinguishers should have the handle at 1 m above floor level for large extinguishers and 1.5 m for smaller ones (up to and including 4 kg). Trolley-mounted extinguishers are an alternative. To safeguard against spinal damage, a greater number of smaller extinguishers might be acceptable, depending on the results of a specific risk and health and safety assessment.

It ought to go without saying that risk assessments for portable fire extinguishers must go hand in hand with a total building risk assessment and a premises evacuation strategy. Hand-held fire extinguishers are intended to be used by suitably trained people – but will inevitably be wielded by untrained staff. Therefore, care should be taken to advise all likely users of the correct use, function and dangers of each type of extinguisher. As the RRO responsibilities – and liabilities – become clearer, whether 'responsible persons' begin to downplay fire extinguishers in favour of building evacuation, remains to be seen.

Note: See also Fire Information Association's FIA Factfiles 0001–0008 for detailed specification on the main types of extinguisher, www.fia.uk.com

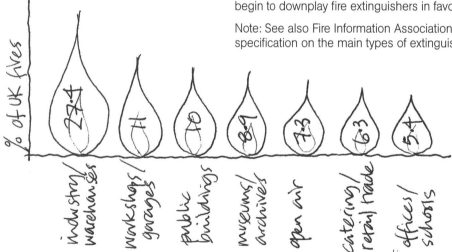

References

BS 7937 (2000) *'Specification for portable fire extinguishers for use on cooking oil fires'*, BSI

BS 5306-8 (2000) *'Fire extinguishing installations and equipment on premises – Part 8: Selection and installation of portable fire extinguishers – Code of practice'*, BSI.

BS EN 2 (1992) *'Classification of fires'*, BSI.

BS EN 3-7 (2004) *'Portable Fire Extinguishers – Part 7: Characteristics, performance requirements and test methods'*, BSI

Chief Fire Officers' Association (2006) *'A Short Guide to Making Your Premises Safe from Fire'*, HM Government.

Health and Safety Executive (2004) *'Getting to Grips with Manual Handling: A short guide'*, HSE.

Health and Safety Executive (1996) *'Health and Safety (Safety Signs and Signals) Regulations 1996'*, HSE.

RECOMMENDED READINGS

BS 9999 (2008) *'Code of practice for fire safety in the design, management and use of buildings'*, BSI.

Department for Communities and Local Government (2007), *'Building Regulations Approved Document Part B: Fire Safety'*, NBS.

Department for Communities and Local Government (2006) *'Guides 1–11'* inclusive (including an *'Entry Level Guide'*), available on www.communities.gov.uk.

Office of the Deputy Prime Minister (2002) *'Building Regulations Approved Document Part B: Fire Safety'*, NBS.

30: Sprinklers Automatic fire protection

It has been estimated that of the half a million fires in the UK every year, only 60,000 of them occur in dwellings, but these cause a disproportionate number (75 per cent) of all fire-related fatalities. Building Regulations Approved Document B believes that sprinkler systems could go some way to reducing these numbers.

After all sorts of internal machinations, accusations and infighting, the Residential Sprinkler Association (RSA) was set up in March 2007, taking over from the liquidated Fire Sprinkler Association (FSA), and separate from the British Automatic Sprinkler Association (BASA) which has changed its name from the British Automatic Fire Sprinkler Association (BAFSA). Then there's the National Fire Sprinkler Network (NFSN), the European Fire Sprinkler Network (EFSN) and the Fire Industry Confederation (FIC).

This liberal sprinkling of professional and trade bodies dealing with this subject all hope to correct the misconceptions about sprinkler installation. One of the reasons for the (re-)formation of the RSA was the need for a coordinated response to the latest edition of Approved Document B: Fire Safety (AD B) which, for the first time, addresses the use of sprinklers versus door-closing devices, introduces a maximum unsprinklered compartment size for single-storey warehouses, improves measures to assist firefighters tackling blazes in tall buildings and introduces a risk-based approach to automatic fire protection.

A sprinkler system is very straightforward: it comprises a water-filled network of pipes laid throughout the building with branches off to sprinkler heads at various key locations. Each sprinkler head is activated by either a fusible link (one that melts) or a heat-sensitive liquid-filled glass bulb (which bursts) when the temperature gets to a certain level. They

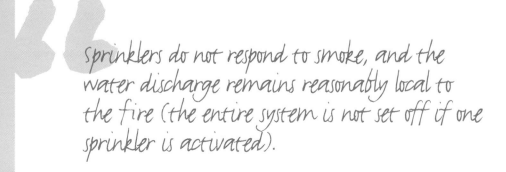

Sprinklers do not respond to smoke, and the water discharge remains reasonably local to the fire (the entire system is not set off if one sprinkler is activated).

do not respond to smoke, and the water discharge remains reasonably local to the fire (the entire system is not set off if one sprinkler is activated). In domestic situations, sprinkler heads are seldom necessary in bathrooms, toilets, wardrobe/cupboard spaces and ceiling cavities as almost 90 per cent of fatal fires originate in bedrooms, lounge/dining rooms and kitchens.

The sprinkler heads come in many shapes and sizes but tend to be one of the following:

- pendent (not spelled 'pendant') sprinklers, where nozzles direct the water downwards

- recessed sprinklers, where the heat-sensing element is above the lower plane of the ceiling

- concealed sprinklers, usually recessed but covered by a plate that detaches and falls away under the influence of heat. Concealed sprinklers are often preferred because of their unobtrusiveness, although their efficiency and usefulness can be compromised if they are painted over

A sprinkler discharges 60 l/min (compared to a fireman's hose that discharges around 600–1000 l/min, with, according to a number of fire services, an average of around 3000 litres required per domestic fire). In general, a sprinkler system will use less than 5 per cent of the water used by the fire service.

Due to uneven water pressures in certain areas, the Department for Communities and Local Government (DCLG) assumes that this sprinkler system water pressure requirement will rule out around 25 per cent of all homes from installing sprinklers. In single dwellings where the incoming supply only feeds the sprinkler system, 60 l/min will normally be accommodated with pipework of at least 25 mm internal diameter, but the pipework also needs to be sized to maintain a flow rate of at least 42 l/min if two

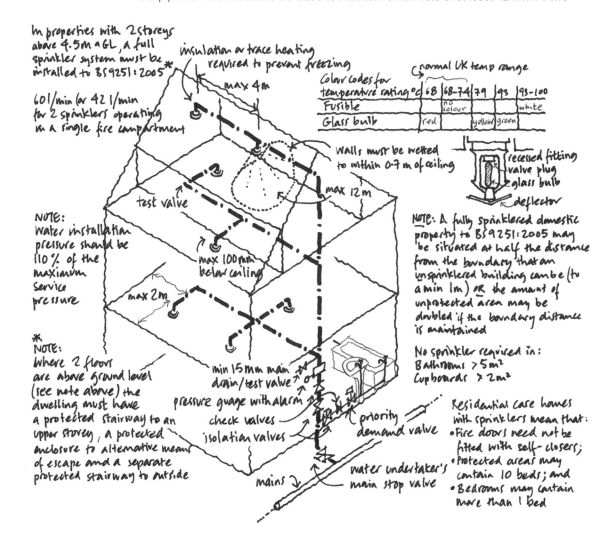

sprinkler heads are activated (42 l/min if *four* sprinkler heads are activated in residential accommodation). In single dwellings where the sprinkler system is connected to the domestic mains water supply, then the design flow rate needs to include an additional 25 l/min (an additional 50 l/min in residential accommodation); this factors in for pressure drops when taps are opened. Check the water supply before proceeding with sprinkler system design, and it is advisable to consult with all of the following parties:

- **The water undertaker** has to ensure that the works comply with Water Supply (Water Fittings) Regulations 1999. There are ten water and sewerage service companies in England and Wales with an additional 15 companies providing water supply only. In Scotland and Northern Ireland, water and sewerage services are provided by single public companies, Scottish Water and the Department for Regional Development's Water Service respectively.

- **The local fire authority** – A warning sign should be fitted near the householder's mains isolation valve to make it clear that isolating the incoming water supply will disable the sprinkler system. Similarly, firefighters should be easily directed to shut off valves and electric cut-off switches.

- **The Building Control body** – If the assessed risk of fire in a non-domestic building is local, partial building coverage can be negotiated with the authorities. Where sprinklers are used to reduce the risk profile, i.e. to increase the travel distance allowances, then sprinklers must be installed in accordance with BS EN 12845 (new systems) or BS 5306-2 (existing systems) but not BS 9251. For example, if the minimum level of fire protection in a building provides the maximum length of one-way travel at 22 m, by fitting sprinklers the risk alters, and the maximum length of one-way travel can be increased to 26 m. BS 9999 suggests that buildings used for the consumption of alcohol should have a higher risk – and hence shorter travel distances – factored in.

  (Note: During the construction of non-domestic premises, consideration should be given to workers involved in fitting out and finishing who may, to all intents and purposes, be occupying a building which has had escape distances assessed by taking the sprinkler system into account. Logically, the sprinkler system should be installed and operational to protect these workers.)

- **The insurer** – The BRE says that 'residential sprinklers are not cost-effective for other dwellings' (see below). An oft-cited article in the online American journal, ScienceDaily, quotes research by the federally funded US National Institute of Standards and Technology that presents a clear economic case for fitting sprinklers in houses. It states that the use of sprinklers actually results in a '32 per cent reduction of both direct property damage (property losses that would not be covered by insurance) and indirect property costs (fire-related expenses such as temporary shelter, missed work, extra food costs, legal expenses, transportation, emotional counselling and childcare)'.[1] It's even more cost-efficient if you factor in the dental costs resulting from insurance broker's gnashing of teeth.

A survey by BRE in 2004 showed that even though residential sprinklers do fulfil a vital function of successfully controlling and extinguishing a fire, pendent-style fittings are generally only cost effective for residential care homes and those dwellings above 11 storeys. Consequently, AD B requires that sprinklers be provided in blocks of flats exceeding 30 m in height and sprinklers in residential care homes mean that fire doors need not be fitted with self-closers, protected areas may contain 10 beds, and bedrooms may contain more than one bed. Cost savings aside, sprinklers are now required in domestic dwellings with two or more storeys at 4.5 m above ground level and in all new schools other than those with 'low risk'.

The BRE report 'Effectiveness of sprinklers in residential premises' revealed that 'slow-growing' and shielded fires are more problematic for sprinkler efficiency. As part of the Regulatory Reform (Fire Safety) Order, owners should be aware that the careless stacking of storage material, say, may cause sprinklers to be partially shielded, thus slowing down their response time and obstructing some of their discharge. Correcting this should be a standard part of the routine maintenance carried out by a competent person whose activities may also include checking of pressure gauges, alarm systems, water supplies, booster pumps, etc.

Where sprinklers are used to reduce the risk profile, i.e. to increase the travel distance allowances, then sprinklers must be installed in accordance with BS EN 12845 (new systems) or BS 5306-2 (existing systems) but not BS 9251

At mid-2007 prices, sprinkler system installation costs range from around £2100 for a fully furnished house, to £1100 for a vacant property and £500 for new-build (although these figures may vary depending on location and expertise, as well as profit margins)[2]. The bulk of the cost of retro-fitting the system in existing properties comes from the installation of a new water supply pipe and mains connection, varying from 40–60 per cent of the total system cost. Maintenance costs are assumed to be around £500 per annum. However, in lofts, for example, especially those in cold roofs, care must be taken to protect the pipework from freezing, either using anti-freeze, or wrapping the pipework in insulation (taking care not to interfere with the sprinkler head), or using electrical trace heating. Systems using anti-freeze must not be directly connected to the mains, and plastic pipes should not be used with glycol-based anti-freeze (glycerine is OK). These can add significant additional costs to the scheme.

BRE's follow-on research paper in March 2006 examined the suitability of concealed and recessed pattern sprinklers for use in residential premises, particularly concerning their effectiveness and maintainability. It revealed that they operated later and at higher temperatures than the exposed pendent sprinklers and were therefore not adjudged to have performed with the same thermal sensitivity rating ('quick response') that was noted for pendent sprinklers. Given that the experiment addresses various types of fire likely to be encountered in residential premises, several key aspects are worth mentioning. Recessed fittings were not adjudged to have performed satisfactorily in response to table fires in terms of saving lives, although they brought the fire under control (they were acceptable in response to television fires). Fittings recessed to the maximum depth permitted by the manufacturer are relatively slow to respond and none of these fittings could be classified as 'quick response', even though they manage to control the fire.

On 2 November 2007, a fire in a large warehouse in Atherstone-on-Stour took 100 firefighters 12 hours to extinguish but left four part-time firefighters dead – the worst loss of life in the British fire service since 1972. Under UK fire legislation there is no requirement to put sprinklers in such warehouses (those up to 20,000 m²). However, there is a clear duty under the Corporate Manslaughter and Corporate Homicide Bill, CDM and Health and Safety legislation to protect and safeguard relevant persons. As revealed in Shortcuts: Book 2, the legislation covers 'general fire precautions' and other fire safety duties which are needed to protect 'relevant persons' in case of fire in and around most 'premises'. Under the legislation, a firefighter is not a relevant person. The premises did not have sprinklers throughout.

[1]'Home Fire Sprinklers Score 'A' In Cost–Benefit Study', Science Daily, 15 Oct 2007 which makes reference to David T. Butry, M. Hayden Brown and Sieglinde K. Fuller, 'Benefit–Cost Analysis of Residential Fire Sprinkler Systems', National Institute of Standards and Technology

[2]These are rounded-up figures from 'Development of a lower-cost sprinkler system for domestic premises in the UK', (April 2007)

References

BRE (1992) Report 213 *'Sprinkler operation and the effect of venting: studies using a zone model'*, BRE.

BRE (2004) Report R 204505 *'Effectiveness of sprinklers in residential premises'*, BRE.

BRE (2006) Report R 218113 *'Effectiveness of sprinklers in residential premises – an evaluation of concealed and recessed pattern sprinkler products'*, BRE.

Defence Estates (1999) Technical Bulletin 99/33 *'Fire prevention and fire safety: inspection, testing and maintenance of sprinkler installations'*, DE.

Loss Prevention Certification Board (2005) Loss Prevention Standard 1039 *'Requirements and testing methods for automatic sprinklers. Issue 5.1 dated 07/11/01'*, LPCB.

Loss Prevention Certification Board (2007) Loss Prevention Standard 1301 *'Requirements for the approval of sprinkler installers in the UK for residential and domestic sprinklers'. Issue 1 dated June 2007*, LPCB.

RECOMMENDED READINGS
BS 9251 (2005) *'Sprinkler systems for residential and domestic occupancies – Code of practice'*, BSI.

BS 9999 (2008). *'Code of practice for fire safety in the design, management and use of buildings'*, BSI

BS EN 12845 (2004) *'Fixed firefighting systems – Automatic sprinkler systems – Design, installation and maintenance'*, BSI.

Fire Protection Association (2007) *'Development of a lower-cost sprinkler system for domestic premises in the UK: Fire Research Technical Report 2/2007'*, April 2007, Department for Communities and Local Government: London.

(Note: At the time of writing, the government advises that no systems should be built in the UK to comply with this report since some of its recommendations 'currently' do not comply with the Water Supply (Water Fittings) Regulations 1999.)

Fire Protection Association (2004) *'Guidelines for the supply of water to fire sprinkler systems'*, FPA.

Fire Protection Association (2005) *'LPC Rules for automatic sprinkler installations – incorporating BS EN 12845'*, FPA.

Loss Prevention Certification Board (2005) Loss Prevention Standard 1036 *'Quality schedule for the certification of automatic fire sprinklers'. Issue 2.1 dated 1 July 1993*, LPCB.

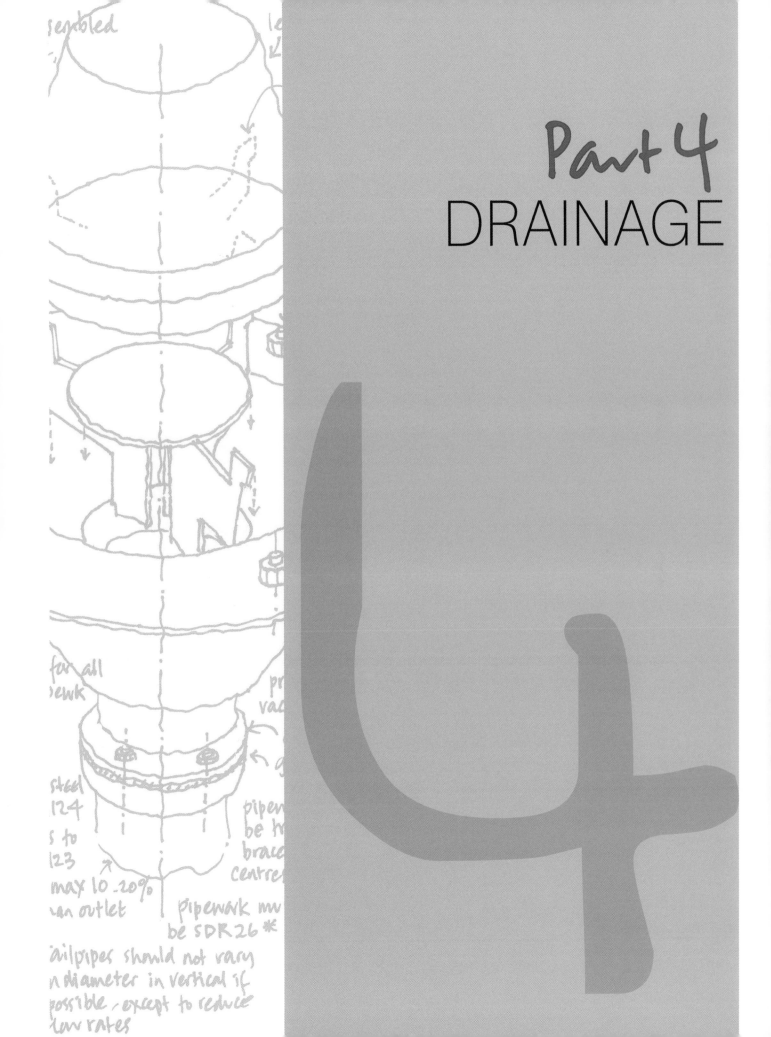

Part 4
DRAINAGE

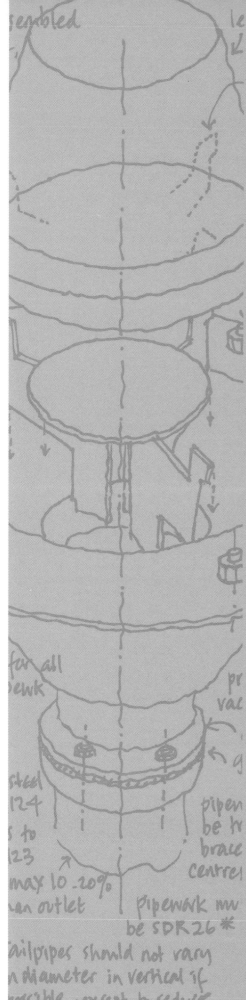

sembled

le

for all
pewk

pr
vac

g

steel
124
s to
123
max 10 .20%
an outlet

piben
be tr
brace
centre

pipework mw
be SDR 26 *

ailpipes should not vary
n diameter in vertical if
possible , except to reduce
low rates

31: SUDS
Sustainable urban drainage systems

With concerns about flooding and soil erosion becoming widespread, what are the options for managing rainwater run-off with minimal reliance on, and investment in, mains infrastructure? Here we examine the technicalities, approvals and consents needed for sustainable urban drainage.

Many of the historic drainage systems that lie beneath our feet combine surface water and foul water in one pipe. After all, the early Victorian combined sewers tended simply to discharge the effluent into a river. As a result of London's Big Stink in 1868 (so vile that Parliament's tactic of perfuming the curtains in order to mask the stench, proved ineffective), pipes were routed further afield – extended downstream (towards the estuary in London's case) and eventually, by the end of the 19th century, into sewage treatment works.

Currently, in east London, when these old combined sewers overload, the excess discharges (intentionally) into the Rivers Lee and Thames to reduce the risk of London's sewage treatment works overflowing and flooding properties. The National Trust *et al.* claim that combined sewers 'discharge 18 million gallons of raw sewage and urban run-off into the Thames every year.'[1] A commission has been set up to rectify this before the 2012 Olympics.

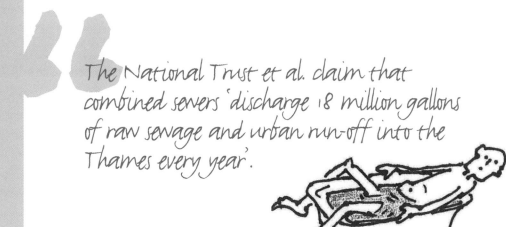

The National Trust et al. claim that combined sewers 'discharge 18 million gallons of raw sewage and urban run-off into the Thames every year'.

Designers of modern developments are generally obliged to separate out sewage and surface water to avoid heavy rainfall placing an intolerable burden on the treatment facilities into which they would otherwise both discharge. An overflow from an overloaded sewer is known as 'hydraulic flooding', meaning that the volume and/or flow rate exceeds the hydraulic capacity of the pipe. Both speed and volume of flows in combined drains are greatly affected by heavy downpours. Floods arising from surface water drains are bad enough, but combined sewer floods don't bear thinking about. Nowadays, foul water is dealt with separately and, where possible, designers are urged to route surface water such that it discharges into natural watercourses or into the soil via controlled infiltration methods.

One important consideration in separate surface water drainage in 'natural' drainage schemes is the need to reduce pollution and damage to the water table. Pollution may derive from the surfaces that have been drained (the contaminated run-off from lead roofs and oily driveways, for example), while damage can be caused by an unmediated rate of flow into the watercourses into which the surface water discharges, leading to scouring and flooding during severe rainstorms. The Environment Agency (EA) normally only requires pollution hazard assessments for lorry parks, garages, industrial sites and major commercial areas although some less hazardous places may fall into its remit. For instance, if the rainwater goods from a domestic roof are not sealed but discharge to an open grating, EA authorisation is required.

Building Regulations Approved Document H: 2002, 'Drainage and Waste Disposal' (AD H), sets out the benchmarks for dealing with rainwater disposal. It states that adequate provision shall be made for rainwater run-off from roofs and paved areas to discharge, in order of priority, into a soakaway, a watercourse or, where these are 'not reasonably practicable', into a sewer.

Proving that something is 'not reasonably practicable' is not easy, and nowadays it is rare to discharge rainwater drainage into a 'non-natural' means of disposal. In essence, when designing drainage systems, soakaways and watercourse connections must be explored and proven to be inappropriate before plans to take rainwater into a run-off sewer will be considered for approval. After all, AD H states that methods of drainage 'other than connection to a public surface water sewer are encouraged where they are technically feasible'. Scottish Planning Policy (SPP) 7 'Planning and Flooding' effectively says the same. Planning Policy Guidance (PPG25) and Planning Policy Wales Technical Advice Note (TAN 15) (both entitled 'Development and Flood Risk') encourage local planning authorities to adopt sustainable drainage 'wherever practicable'.

Recent floods in the UK seem to have occurred after non-exceptional bouts of rainfall and have given some people cause for concern. Even though impact tests are regularly required, rainwater is often casually discharged into long-defunct drainage channels or expected to naturally filtrate into poor absorptive soil, so it is hardly surprising that it sometimes cannot cope. Phil Rothwell, head of flood management at the Environment Agency (EA) has said that, with limited investment in new piped infrastructure, the UK's drainage system is 'failing'.

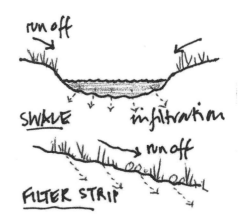

The objective of SUDS is to manage water run-off rates, but equally to provide habitat for wildlife, protect water quality and, where appropriate, to encourage natural groundwater recharge. To this end, local controls on run-off quantities and qualities have to be managed close to source. Therefore, CIRIA advises that, where possible, run-off be limited by recycling at source, using water butts, incorporating green roof design, etc. After all, the EA reports that for every 1 per cent reduction in run-off surfaces there will be a 9 per cent decrease in sewer-related flooding. Under CIRIA's water rationing agenda, it also suggests that pollution can be kept to a minimum 'by keeping paved areas clean' (though not by washing, but by 'sweeping hard surfaces regularly'). Other helpful tips include the memorable: 'Litter and animal faeces can be kept out of drainage systems by education and the provision of bins'. A number of Housing Associations now insist on low volume baths to limit water use and drainage 'strain' at source.

Given the legislative and practical constraints, sustainable urban drainage (SUDS) systems use a variety of techniques to control surface water run-off locally, avoiding or considerably reducing these problems.

FILTER STRIPS AND SWALES
Filter strips are areas of ground, covered in vegetation, that absorb run-off from adjacent hard surfaces. Swales resemble long shallow ditches, and can have the additional function of carrying the collected water elsewhere. These features slow down the flow of surface water, and remove pollutants through the filtering effect of the planting, which may be slow-growing grasses or wild flower mixtures. Gradients must be gentle to avoid erosion, and deep hollows in which planting may become waterlogged should be avoided.

PERMEABLE SURFACES OR PAVING
Paving is traditionally laid to falls to gullies or slot drainage, with below-ground pipework to sewers. In a SUDS scheme, the run-off can be treated and disposed of locally, e.g. via soakaways. A more radical solution is to use permeable paving, where the use of below-ground pipework can be eliminated or much reduced.

SOAKAWAYS, INFILTRATION TRENCHES AND FILTER DRAINS
These features all include some degree of water storage which is designed to delay the release of water to the sewerage system, or into the surrounding ground. Specific details on soakaways are included in this section but, in general, trenches usually rely on an excavated pit or trench filled with coarse aggregate or rubble often with a geotextile lining and sometimes with a precast concrete or brick structure to retain the fill. Vertical pipes capped by inspection covers may be used to monitor water levels and assist maintenance. A filter drain is essentially an infiltration trench with a horizontal pipe to assist water flow. Alternatively, plastic honeycomb units, again wrapped in geotextile, can be used to create underground voids where water can be held before percolating out.

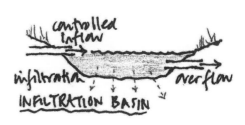

BASINS AND PONDS
These store water above ground, temporarily or permanently just like a pond. Flood plains and detention basins are normally dry but flood harmlessly during heavy rainfall. 'Balancing' and 'attenuation' ponds are permanent water features that moderate the flow

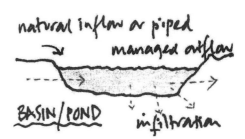

Rainwater is often casually discharged into long-defunct drainage channels or expected to naturally filtrate into poor absorptive soil, so it is hardly surprising that it sometimes cannot cope. Is that the cause of increased flooding?

by storing the run-off into the basin (or pond) during flood conditions and releasing it after the peak has passed. 'Retention' ponds retain the water for longer periods to improve quality, by allowing pollutants to settle or by active biological treatment of the water. With the addition of specific planting to form constructed wetlands, the quality improvement is even greater, allowing secondary or tertiary treatment of effluent from septic tanks or sewage treatment units. Basin and pond linings can be of compacted clay or bentonite, or sheet material (including geosynthetic clay).

REGULATORY FRAMEWORK

Because SUDS is a new and developing phenomenon and departs radically from traditional 'pipe away and treat remotely' systems, the permissions and approvals necessary as the project evolves may be unfamiliar and conflicting. For example, planning permission may be needed for some features although regional planning guidance (or regional spatial strategies) will tend only to have generic requirements. The more specific requirements will be spelled out in structure plans, local plans and supplementary planning guidance (SPGs). These tend to require the applicant/agent to demonstrate that, where SUDS is not practicable, a full environmental impact report is compiled, including such details as 'the aims of the SUDS approach to drainage' and 'local soil and hydrology characteristics'. In Scotland, the Drainage Impact Assessment (DIA) requires developers to document the impact of the proposed development on the catchment, including soil porosity and classification data and proof of attenuation for critical rainfall periods.

Building Regulation approval is needed where a SUDS system is used for run-off from building roofs and paving associated with buildings. The following agency approvals may be required:

- sewerage undertaker for connections to adopted sewers, and sewerage systems that are to be adopted

- highway authority for highway construction and drainage

- local Authority where public land is to be maintained

- environmental regulator (e.g. Environment Agency, Scottish Environmental Protection Agency) for consent to discharge, e.g. to a watercourse

The final thing to note is that, unlike most correctly installed piped drainage systems, sustainable urban drainage systems require close monitoring and maintenance, from regular de-silting and the disposal of sediment to grass cutting. This needs to be factored in at an early stage. Some councils will adopt publicly accessible systems under Section 106 Agreements on the provision of a maintenance plan and requisite payment arrangements. In some cases, adoption by the statutory sewerage undertaker may be possible, although it has no duty to accept the land drainage flow.

[1] 'Blueprint for Water: 10 Steps to Sustainable Water by 2015', www.blueprintforwater.org.uk

References

Construction Industry Research and Information Association (2007) 'The SUDS manual', CIRIA.

Construction Industry Research and Information Association (2004) Publication C625 'Model agreements for sustainable water management systems. Model agreements for SUDS', CIRIA.

Construction Industry Research and Information Association (1995) Report R 134 'Sediment management in urban drainage catchments', CIRIA.

Department for Communities and Local Government (2006) 'Planning policy statement 25: development and flood risk – full regulatory impact assessment', DCLG.

Department for Communities and Local Government (2008) 'Planning policy statement 25: development and flood risk – practice guide, DCLG.

Environment Agency (2003) 'Sustainable drainage systems (SUDS) – A guide for developers', EA.

Planning Policy Wales (2004) Technical Advice Note (TAN) 15 'Development and Flood Risk', Welsh Assembly Government.

Scottish Executive Development Department (2001) Planning Advice Note 61 'Planning and sustainable urban drainage systems', SEDD.

Scottish Executive Design Development (2004) Scottish Planning Policy (SPP) 7 'Planning and Flooding', SEDD.

RECOMMENDED READINGS
Construction Industry Research and Information Association (2006) Publication C635 'Designing for exceedance in urban drainage – good practice', CIRIA.

National SUDS Working Group (2004) 'Interim Code of Practice for Sustainable Drainage Systems', CIRIA.

Office of the Deputy Prime Minister (2002) 'Approved Document Part H: Drainage and waste disposal', ODPM.

32: Roof Drainage Sucks
An overview of siphonic rainwater systems

If you believe the hype, it's cheaper, quicker, more effective, minimises excavations, increases useable internal areas, reduces loads, optimises voids and even ticks the boxes of sustainability, innovation, efficiency and 'maintenance free'. So why is it that siphonic drainage still has only a marginal market in the UK?

The first patent for siphonic drainage was granted in Scandinavia in 1968, with its first major application on a Swedish turbine factory in 1972. It took a further 20 years or so before siphonic (also spelt 'syphonic') systems were tried out in the UK. Even though they have, to a certain extent, proved their worth, architects and clients are still uncertain about their merits. An academic report written in 2001, for example, could find just one building in America that used siphonic drainage.

So what are the differences between conventional and siphonic roof drainage systems? The former generally comprises external or internal gutters connecting to vertical downpipes at regular intervals. On large flat roof areas draining away from the external wall, roof outlets may connect with internal downpipes, or there may be internal gutters that fall to the perimeter. Usually, high-level connections to internal downpipes are laid to falls and each individual downpipe connects to the underground drainage system. This results in deep ceiling voids to accommodate the pipe falls, downpipes breaking up the floor plate, and considerable trenching under the slab as well as around the perimeter.

Flow velocities within the pipework must be kept above a certain minimum to prevent air infiltration into the system (known as 'cavitation') resulting in lower pressure differentials and consequent flow disruption.

Whatever the layout, the flow in the downpipes (as anyone au fait with Mitchell's construction manuals will know) is 'annular', meaning that the water corkscrews around the internal surface of the pipework with a central column of air at the centre. This flow condition needs relatively low flow velocities and large diameter downpipes (typically 100–150 mm). The dimensions, airflows and falls ensure a suitable overall capacity for self-cleansing flow velocities.

Conversely, siphonic roof drainage is a nominally closed system. By restricting the amount of air getting into the system, a full bore water flow condition is created which actually pressurises the pipework. A typical siphonic system comprises a series of flush-mounted outlets laid out across a given roof area (along the flat roof or within the gutter). These look similar in size and shape to 'conventional' outlets, but incorporate baffles restricting the amount of air that gets in to prevent vortices forming. As the water enters the system (the pipework has a minimum internal diameter of 32 mm) and the volume and flow build up, the air is purged (called 'priming' the system) and siphonic action occurs. This means that the flowing column of water creates negative pressure behind it, drawing more water in.

FLOW RATES

Undoubtedly, siphonic drainage systems need more computational modelling and engineering calculations than for a traditional gravity discharge system, and back-of-a-fag-packet calculations are not recommended. For example, flow velocities within the pipework must be kept above a certain minimum to prevent air infiltration into the system (known as 'cavitation') which would result in lower pressure differentials and consequent flow disruption. Also, consideration must be given to the layout since, if the height of the building and hence the downpipe is too great, then the speed of flow becomes excessive and inappropriate negative pressures can build up leading to potential failure of the whole system. In buildings where this might happen, downpipes can incorporate a section with enlarged diameter to slow the flow without introducing air; effectively making a portion of the downpipe into a conventional gravity-fed, annular flow pipe.

In siphonic systems, high-level pipework can be run horizontally, rather than to falls, thus reducing the ceiling void depths required. There are generally fewer outlets, which, in turn, require even fewer downpipes resulting in less complicated below-ground drainage. Self-cleansing velocities are easier to achieve than with conventional systems and, as a consequence, rodding/access points are seldom required.

The siphonic system should terminate in a vented main sewer/drain (vented to prevent the risk of surcharging) or discharge straight into a water-retaining chamber/soakaway/open drain that has been suitably sized for the volume and flow rate. If the rainwater outlet discharges into a sustainable urban drainage system (SUDS) then the filtration trench or pond can be less deep than those associated with conventional drainage systems because the shorter runs of underground drainage associated with siphonic systems lead to shallower inverts. However, bear in mind that the peak flow tends to be greater in a siphonic system thus requiring greater SUDS capacity.

High density polyethylene (HDPE) is the preferred pipework material for this type of system because of its workability and wide range of sizes. Connections can be made by fusing pipes together. Sleeve fusion jointing of HDPE pipework is recommended, whereas on-site butt-jointing should only be permitted if 'factory conditions' can be guaranteed. Metal pipework can also be used although the internal roughness quotient is part of the flow calculation for an effective system and metal frequently causes higher frictional losses than HDPE and other plastics.

LARGE AND SMALL AREAS

Predominantly, siphonic drainage is suited to the traditional large surface area roofs normally associated with the retail, assembly, manufacturing and commercial sector as opposed to the smaller areas of the domestic and residential markets. Chep Lap Kok Airport in Hong Kong and Sydney's Olympic Stadium have used it to good effect as have Barajas Airport in Madrid, HM Treasury and the GLA Building. As a rule of thumb (notwithstanding the specifics of the building under consideration) build costs can be reduced by using siphonic systems on buildings over 1500 m² (some manufacturers say 3000 m² or over).

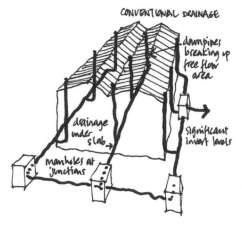

CONVENTIONAL DRAINAGE

downpipes breaking up free flow area

drainage under slab

significant invert levels

manholes at junctions

SIPHONIC DRAINAGE

siphonic outlet

tailpipe

collector pipe

collector pipe

downpipe

manhole

shallow invert

Most systems have limited effectiveness on small areas (balconies, for example), where the rainfall rate is difficult to calculate accurately. These small areas could possibly disrupt the general flow of the overall system and ought to be kept separate from the siphonic system. When calculating the number and location of outlets for large areas, the flow rate calculation criteria are:

- internal diameter of the pipe work (or bore)

- number and configuration of outlets

- specific roof flow conditions

- material(s) used for the pipework

- actual local rainfall intensity rate.

The last figure can be read off one of the five different maps in BS EN 12056-3, dependent on the risk factor ranging from 2-minute storm intensities with a return factor of one year, all the way up to 500 years. The technical team at drainage company, Geberit, notes that: '108 mm/hr has proved to work well in most locations in the UK with several 2-minute occurrences annually'. Even though the BS EN 12056-3 figures need to be rounded up to suit the rainwater run-off calculations, designers may feel tempted to introduce an extra contingency factor to take account of the increasingly litigious attitude of the insurance industry's global warming fears that have increased since these figures were written in 2000. Note: This temptation should be resisted – or treated with caution – as flow data must not be arbitrarily plucked out of the air. Pipework *must* be designed for actual rainfall rates and an honest appraisal is essential to ensure the proper operation of the system.

Heriot-Watt University is carrying out a range of other significant research exercises on siphonic theory and practice. It says, with some candour, that 'steady-state design

Rainfall intensity must be determined by reference to BS EN 12056:3

NOTE: If disassembled for maintenance, all parts must be re-assembled.

leaf guard

dotted lines indicate the position of the baffle vanes when leaf guard is in place

Blockages in an outlet can lead to cavitation elsewhere in system

gutter design to BS EN 12056:3

baffle vanes

fixing bolts

outer bowl

anti-vortex plate restricts the flow of water and air to ensure full bore characteristics

Avoid contact of dissimilar materials that might give rise to electrolytic corrosion

Once the system is "primed" the fall of water sucks the rest behind

Ensure max siphonic flow rate of 1m/sec

Minimal operational pressure of 7-8m for all non-HDPE pipewk

non-vertical pipewk to be horizontal where possible

minimal operational pressure of 8-8m vacuum for HDPE

·Cast iron to BS EN 877

connector

gasket seal

·Stainless steel to BS EN 1124

pipework must be transversely braced at 15 m centres max

·Galv ms to BS EN 1123

Tailpipe max 10-20% wider than outlet

pipework must be SDR 26 *

·Tailpipes should not vary in diameter in vertical if possible, except to reduce flow rates

·Increasing diameter of collector (horizontal) pipes should be eccentric so top edge is constant

* SDR is the 'standard dimensional ratio' of outer pipe diameter divided by the pipe wall thickness. SDR 26, for example, means that the diameter is 26 times bigger than the pipe wall thickness.

> *The life of the building should be calculated on the basis of more than 30 years (including the safety factor) otherwise, the Siphonic Roof Drainage Association predict that 'there is a serious risk that water will flow into the building during its lifespan'.*

methods are not applicable when a siphonic system is exposed to a rainfall event below the design criteria, or an event with time-varying rainfall intensity'. This really means 'light', or 'intermittent' rainfall patterns, which actually, it admits 'are the norm'. It continues, 'problems are exacerbated when the system incorporates more than one outlet connected to a single downpipe ... as the breaking of full-bore conditions at one outlet (due to low gutter depths and air entry) is transmitted throughout the system and, irrespective of the gutter depths above the remaining outlet(s), results in cessation of fully siphonic conditions.'

Arthur and Wright's 2006 paper 'Siphonic roof drainage systems – priming focused design' states that 'to allow for some limited blockage of the outlet, it is proposed that the design flows should be increased by 10%'.

Consideration should be given to thermally and acoustically insulating the pipework runs. Once a system has been installed, it should not be extended or re-routed without serious thought being given to the implications for the operational integrity of the revised system. In this instance, recalculation will be essential, and expert guidance is recommended.

The Siphonic Drainage Association was set up in 2004 to increase the understanding and specification of these systems and to push for increased standardisation. In March 2007, a new British Standard, BS 8490: 2007 'Guide to siphonic roof drainage systems' was published to help explain the design, installation and maintenance requirements of siphonic drainage systems. In BS-speak, its recommendations do not apply to 'systems using rectangular or trapezoidal section conduits for water conveyance', but only relates to circular-section systems. For other pipework sections, and for more detailed guidance, reference should still be made back to BS EN 12056-3: 2000.

NOTE:

*   The protection life of the building should be calculated on the basis of more than 30 years (including the safety factor) otherwise, the Siphonic Roof Drainage Association predict that 'there is a serious risk that water will flow into the building during its lifespan'.

References

Arthur, S. & Wright, G.B. (2006) *'Siphonic roof drainage systems – priming focused design'*, School of the Built Environment, Heriot-Watt University, Science Direct.

Arthur, S., Swaffield, J.A. & Wright, G.B. (2001) *'Investigation into the Performance Characteristics of Multi-Outlet Siphonic Roof: Drainage Systems'*, Sustainable Water Management Research Group, Heriot-Watt University.

BS EN 12056-1 (2000) *'Gravity drainage systems inside buildings. General and performance requirements'*, BSI.

Building Research Establishment (2000) *Good Building Guide 38 'Disposing of rainwater'*, BRE.

CIRIA (2004) *Publication C626 'Model agreements for sustainable water management systems. Model agreement for rainwater and greywater use systems'*, CIRIA.

Environment Agency (2003) *'Harvesting rainwater for domestic uses: an information guide'*, EA.

Environment Agency (2001) *Conserving Water in Buildings 4 'Rainwater reuse'*, EA.

RECOMMENDED READINGS

BS 8490 (2007) *'Guide to siphonic roof drainage systems'*, BSI.

BS EN 12056-3 (2000) *'Gravity drainage systems inside buildings. Roof drainage, layout and calculation'* (AMD 17041), BSI.

BS EN 13564-1 (2002) *'Anti-flooding devices for buildings. Requirements'*, BSI.

Department for Communities and Local Government (2006) *'Code for Sustainable Homes: A step-change in sustainable home building practice'*, DCLG.

National Federation of Roofing Contractors (2002) *'Roof drainage tables'*, NFRC.

Siphonic Roof Drainage Association guidance. Available at: http://www.siphonic-roof-drainage.co.uk/

33: Anti-flood Drainage
Building (or not) in 'inappropriate' locations

After Sir Nicholas Stern's famous review 'The Economics of Climate Change', the government responded by issuing guidance on better flood defences. Stern predicted that annual flood losses could cost 0.2–0.4% of GDP in 100 years. In the interim, fitting an anti-flood device for a hundred quid or so seems like a pragmatic solution.

Many domestic drainage systems are laid with reasonably shallow falls in compliance with the recommendations in Approved Document H (AD H). This suggests that 75 mm and 100 mm rainwater drains should be laid at not less than 1:100 falls, and 150 mm drains and sewers should be laid at gradients not less than 1:150. The falls on foul drainage depend on the number of connections that affect the flow rate so that a 100 mm diameter pipe with a flow of more than 1 l/sec must be laid at a gradient of not less than 1:80.

In general then, an average small housing estate constructed with minimum drainage falls is likely to have just half a metre of rise from the invert of the main sewer connection to the invert of the shallowest manhole, and possibly only a further 450 mm between that manhole invert and a ground-floor WC trap. If the main combined drain should become blocked, theoretically there is only an arm's length preventing your toilet spilling over.

In severe floods, a storm sewer has the potential to surcharge and at some critical point to overflow.

grey water discharge

removable grille

manual closing override

float allows normal drainage flow but closes off with backflow

water trap

TYPICAL TYPE 2 ANTI-FLOODING DEVICE

NB: TYPE 2 is not suitable for foul water conditions – TYPE 3 for foul water tends to require a motor with 24hr backup

to main drain →

NB: Consult Environment Agency (E+W); Dept of Environment (NI) and Scottish Environment Agency (SCOTLAND) and local authorities for site specific advice

BS 1717: 2001 provides detailed examples of airlocks and vents to be used in different circumstances to prevent potable water becoming contaminated by backflow. Category 1 water (potable) can be polluted by infiltration of Category 2 (stagnant water, say, from a water tank) through to Category 5 (sewer waste or swimming pool water). Essentially, if a potable water network (including a 'water draining device') connects with a drain, there must be either a physical separation of the supply and the drain 'connections', or a connecting device that has a vent/grille sufficient to ensure that any potential contaminating backflow will overspill rather than back up. AD H states that this is the basic line of defence against contamination in relatively low-risk areas.

HIGH-RISK CONDITIONS

In severe floods, a storm sewer has the potential to surcharge, i.e. the water level in the manhole rises above the top of the pipe, and at some critical point, hydraulic pressure will cause drains below the rising water level to overflow.

The potential for – and fear of – drainage backflow and consequent flooding has been greatly increased of late due to:

- WCs situated in basements
- houses built in low lying/flood-risk areas
- contemporary fears about increased rainfall detrimentally affecting existing combined sewers.

In high-risk areas, AD H recommends anti-flooding valves that can be fitted to the main drains to resist any backflow pressure. These valves comprise a 'gate', like a float valve, that allows drainage water to flow in the designed direction but will automatically cut off if water starts to flow – above a certain volume – in the opposite direction. AD H states that these devices should be of the double valve type, be suitable for foul water, have a manual fail-safe and comply with pr EN 13564.

The AD H description is equivalent to pr EN 13564-1's Type 2 category of anti-flooding device, although, at the time of writing, this is deceptive. Only ACO manufactures an anti-flooding device that is designated to deal with foul water flow and this is a higher specification (Type 3). This is an *automated* device rather than an 'automatic' one. ACO's Type 3 device has a high-torque electric motor with battery backup ensuring that the valve can close even in the presence of solid waste. Such faecal matter would undoubtedly clog up the 'passive' float or gate valve of a Type 2 anti-flooding system. Even though pr EN 13564-1 states that passive Type 2 anti-flooding devices are used in foul drainage systems in Austria, there are no official UK regulatory guidelines on this matter. Therefore, only Type 3 (and possibly a higher specification) is suitable for foul water use. However, note that while Type 2 anti-flood devices cost around £100, Type 3 devices cost around ten times that amount and therefore careful consideration should be given to where Type 3 devices should be fitted.

EVEN HIGHER RISKS

The new Planning Policy Statement 25 'Development and Flood Risk' (PPS25) – superseding Planning Policy Guidance 25 (PPG25) – was released in December 2006 by the Department for Communities and Local Government (DCLG). It extends greater planning powers to the Environment Agency (which is now a statutory consultee in areas of flood risk) and prioritises design risk assessments when considering the viability of construction projects. It also recommends that planning applications are backed up with site-specific flood risk assessments (FRAs) and that local planning authorities give priority to proposals with sustainable urban drainage (SUDS) or other methods of managing the risk of flooding.

Even though PPS25 promotes SUDS run-off and absorption schemes (see pages 129–32) it says that where there is a high likelihood of flooding (a 1 in 100 annual probability) developers should consider creating or restoring 'functional floodplains' (those areas with a maximum 1 in 20 annual probability) in order to 'create space for flooding to occur'.

To assess whether to release land, local planning authorities should carry out a sequential test to show that there are no appropriate and reasonably available sites in areas with a lower probability of flooding. Where this cannot be shown, the exception test will be applied but is only relevant to large sites where the local authority deems the development necessary and there are no other options. Given the tenor of PPS25, coastal and riverside sites will undoubtedly come under more stringent review than more landlocked, inland ones. Currently, information on areas of concern is shown on the Environment Agency's (EA) Flood Map[1], but also can be sourced from Intermap's IFSAR (Interferometric Synthetic Aperture Radar) technology[2]. Where strategic flood risk assessments (SFRAs) are not available, sequential tests will be based on the EA flood map.

FLOOD ZONES IDENTIFYING APPROPRIATE USES & FRA CONSIDERATIONS

	ZONE 1	ZONE 2	ZONE 3a	ZONE 3b
ANNUAL FLOOD RISK	Low risk $<1:1000$	Medium risk River $1:100 - 1:1000$ Sea $1:200 - 1:1000$	High risk River $>1:100$ Sea $>1:200$	High risk $>1:20$
APPROPRIATE USE	All	Hospitals, residential dwellings, nightclubs/bars, non-residential educational, landfill sites, shops, cafes, general industry, non-hazardous waste management. NOTE: Police stations, basements, mobile homes + other highly vulnerable structures will only be approved on passing the Exception Test (identifying sustainability benefits, etc)	Shops, cafes, general industry, non-hazardous waste man. NOTE: Hospitals, residential dwellings, nightclubs/bars, non-residential educational, landfill sites, etc should only be considered under a sequential Test - demonstrating that no reasonably available sites exist in areas of lower flood risk.	Flood control infrastructure, pumping stations, docks + marinas, MOD buildings, coastguard/lifeguard buildings, amenity spaces, essential ancillary accommodation for staff engaged in any of the above.
FLOOD RISK ASSESSMENT CONSIDERATIONS	Sites greater than 1 hectare should briefly explore the risks from other areas and the effect of hard surfaces, etc.. The FRA should include considerations of the possible effects of climate change.	The minimum proportionate requirements. Assess the vulnerability of users, quantify flood risks, assess absorptive capacity of soil. Appraise historic data + outline the positive and negative consequences of the flood risk assessment. NB: Average rainfall and peak river flow is predicted to double in 20 years.	As ZONE 2 but with more detail included: • Assess the ability to reduce the overall flood risk • Consider relocating to another site and reasons for decisions • Ensure that any open space is safeguarded for possible floodwater storage.	A detailed FRA is required as ZONE 3a

> *Defra, the Department for The Environment, Food and Rural Affairs claims that avoiding 'inappropriate development' is the best way to minimise flood risks.*

Controversially, maybe, there is no statutory duty on the government to protect land or property against flooding; that onus rests with the landowner, and individual property owners have a duty not to drain their land such that it has a detrimental effect on that of their neighbours. While the local authority will assess the notional risks in relation to their development plans, it is the responsibility of the developer/designer to comply with a satisfactory FRA 'identifying opportunities to reduce flood risk' and the 'wider sustainability benefits', outlined in PPS1: 'Delivering Sustainable Development'.

Under the new planning regime, risky developments should aim to include 'drainage pathways' to remove floodwater from a site before any significant harm is caused. However, the Department for the Environment, Food and Rural Affairs (DEFRA) claims that the best way to minimise flood risk is to avoid so-called 'inappropriate development'. Thus instead of overcoming natural problems through a combination of design and technology, DEFRA wants to use the ubiquitous 'hazard avoidance' principle instead.

These different approaches have very different consequences. In Holland, the dramatic, 3km long, €2.5 billion Oosterschelde storm surge barrier has been designed to confront the Netherlands' precarious relationship with nature and to reduce, by a massive piece of infrastructure, the risk of flooding to once every 4000 years, freeing up land on which to build. Conversely, in London, fears that the 20-year-old Thames Barrier may soon become obsolete have resulted in a less ambitious suggestion that we avoid building on low-lying land in the first place.

Ironically, the DCLG says that the risk-based PPS25 has been introduced to counter the de facto 'precautionary approach' of the Environment Agency and insurance brokers, whose attitudes would otherwise be 'highly restrictive of new house-building'. Using a risk-based approach to counter the precautionary principle seems counter intuitive, but presumably we'll have to judge its success on whether more land is freed up for development.

[1] The flood map can be viewed on
www.environment-agency.gov.uk
[2] www.intermap.com

References

BS EN 1717 (2001) *'Protection against pollution of potable water in water installations and general requirements of devices to prevent pollution by backflow'*, BSI.

BS EN 13564-1 (2002) *'Anti-flooding devices for buildings – Part 1: Requirements'*, BSI.

Department of Communities and Local Government (2006) Planning Policy Statement 1 *'Delivering Sustainable Development'*, DCLG.

Department of Communities and Local Government (2006) Planning Policy Statement 25 *'Development and Flood Risk'*, DCLG.

HM Treasury (2006) *'Stern review: economics of climate change'*, HMT.

Office of the Deputy Prime Minister (2002) *'Approved Document H: Drainage and waste disposal'*, ODPM.

RECOMMENDED READINGS
Balmforth, D., Butler, D., Digman, C. & Kellagher, R. (2006) C635 *'Designing for exceedance in urban drainage – good practice'*, CIRIA.

34: The Limestone Seeps Tonight
A soakaway, a soakaway, a soakaway ...

To continue the musical theme, the classic 1962 Bernard Cribbins song, famously produced by George Martin, went: 'There I was, digging this hole / Hole in the ground, so big and sort of round it was / And there was I, digging it deep / It was flat at the bottom and the sides were steep.' Essentially, an early prescriptive specification for a soakaway.

Soakaways are glorified holes in the ground into which stormwater is discharged and from which it is gradually allowed to percolate into the surrounding soil. It is essential that a soakaway be sized accurately to make sure that it doesn't overflow (meaning that it has too little capacity for predicted levels of rainwater run-off) or that water is retained within the soakaway structure itself (meaning that the prevailing soil conditions have not been sufficiently assessed to ensure that they facilitate the effective dissipation of water into the surrounding area).

We have already looked at sustainable urban drainage systems (SUDs), pointing out that local and national policy tends to direct designers and developers to drain surface water to a natural watercourse or other infiltration system rather than to a sewerage system. Building Regulations Approved Document H: 'Drainage and waste disposal' states that methods of drainage 'other than connection to a public surface water sewer are encouraged where they are technically feasible'.

Approved Document H states that methods of drainage 'other than connection to a public surface water sewer are encouraged where they are technically feasible'.

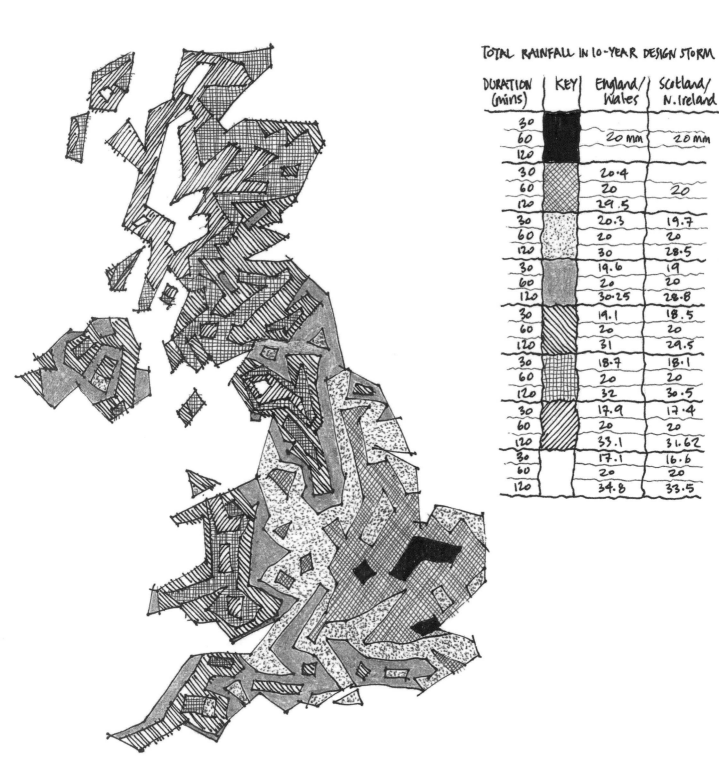

TOTAL RAINFALL IN 10-YEAR DESIGN STORM

DURATION (mins)	KEY	England/ Wales	Scotland/ N.Ireland
30			
60	■	20 mm	20 mm
120			
30		20·4	
60	▦	20	20
120		29·5	
30		20·3	19·7
60		20	20
120		30	28·5
30		19·6	19
60		20	20
120		30·25	28·8
30		19·1	18·5
60		20	20
120		31	29·5
30		18·7	18·1
60		20	20
120		32	30·5
30		17·9	17·4
60		20	20
120		33·1	31·62
30		17·1	16·6
60		20	20
120		34·8	33·5

"To assess whether a soakaway is viable, a site assessment needs to be carried out and a long-winded (but relatively simple) calculation needs to be completed.

The equivalent Scottish Technical Handbook 'Domestic Environment' says pretty much the same sort of thing, with Section 3.6 adding that soakaways in its jurisdiction should be located 'at least 5 m from a building and from a boundary' to ensure that there is no adverse effect on the foundations of the property served by the soakaway or to any neighbouring land. This rule is also contained in Planning Policy Guidance Note 14 Annex 1 and NHBC guidance. Ideally, soakaways should be situated on land that slopes away from the building and also away from other soakaways in the vicinity, so that there is no percolation from one to another. Northern Ireland's Technical Booklet N 'Drainage' was written way back in 1990 and does not contain any details for soakaway construction other than Section N3 requiring that rainwater drains be tested for watertightness, and have adequate falls, flow rates and diameters. Wherever possible, it is advisable to design to the good practice guidance contained within BRE Digest 365 (revised March 2007). Soakaway design software is also available from the BRE.

When planning surface water drainage layouts, it is important to realise that discharge to a main combined drain is becoming less and less acceptable and even discharging to a watercourse will undoubtedly require consent from the Environment Agency. Such approvals may contain onerous provisos for limits on the rate of flow, to ensure that the watercourse is not overwhelmed during heavy rainfall. This will necessitate a flow attenuation device, such as a detention basin, in which run-off can be stored temporarily (and which can also facilitate the biological treatment of pollutants if necessary) before discharging into the watercourse. However, where this type of flow intervention is not practicable, and where a sewer connection is inappropriate, soakaways (and other infiltration mechanisms such as filter drains, swales, etc.) provide a possible solution, although a careful environmental impact assessment must be carried out to avoid overwhelming low-lying/flood plain risk areas.

Of course, soakaways are not always possible. For example, they should not be built in ground where the water table reaches the bottom of the device at any time of the year, or where the presence of contamination within the run-off could result in pollution of naturally occurring groundwater. To assess whether a soakaway is viable, a site assessment needs to be carried out and a long-winded (but relatively simple) calculation needs to be completed. To this end, a trial hole needs to be dug to test the actual infiltration rate into the soil (see diagram). The hole should replicate the proportions of the proposed soakaway (although the NHBC is happy to have narrow boreholes provided that the rate of flow is averaged out over various depths and times). The time taken for water to seep into the surrounding ground needs to be taken for several rapid succession trial runs. Observations – preferably at different times of the day and year – and negotiations with your friendly Building Control officer are needed to confirm that the water table is suitably low.

These percolation tests should be carried out to determine the capacity of the soil (see Approved Document H (AD H2), paragraphs 1.34–1.38). Where the test is carried out in accordance with AD H2, the soil infiltration rate (f) in litres is related to the percolation value (V_p) in seconds/mm) derived from the test by the equation:

$$f = \frac{10^{-3}}{2V_p}$$

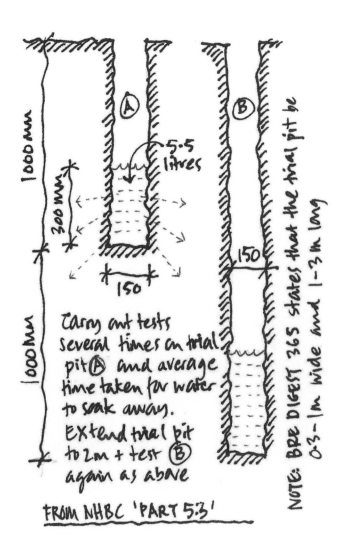

Carry out tests several times on trial pit Ⓐ and average time taken for water to soak away. Extend trial pit to 2m + test Ⓑ again as above

FROM NHBC 'PART 5.3'

NOTE: BRE DIGEST 365 states that the trial pit be 0.3–1m wide and 1–3m long

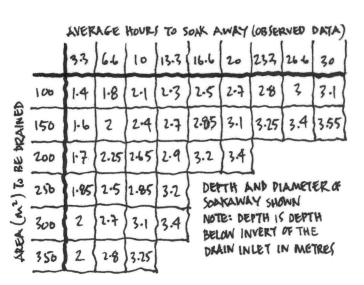

AREA (m²) TO BE DRAINED	\multicolumn{9}{c}{AVERAGE HOURS TO SOAK AWAY (OBSERVED DATA)}
	3.3	6.6	10	13.3	16.6	20	23.3	26.6	30
100	1.4	1.8	2.1	2.3	2.5	2.7	2.8	3	3.1
150	1.6	2	2.4	2.7	2.85	3.1	3.25	3.4	3.55
200	1.7	2.25	2.65	2.9	3.2	3.4			
250	1.85	2.5	2.85	3.2					
300	2	2.7	3.1	3.4					
350	2	2.8	3.25						

DEPTH AND DIAMETER OF SOAKAWAY SHOWN

NOTE: DEPTH IS DEPTH BELOW INVERT OF THE DRAIN INLET IN METRES

The storage volume of a soakaway should be calculated so that, over the duration of a heavy storm, it is sufficient to contain the difference between the inflow volume and the outflow volume. The inflow volume is calculated from the rainfall depth and the area drained. The outflow volume (O) is calculated from the equation:

$$O = as50 \times f \times D$$

where as50 is the area in m^2 of the side of the storage volume when filled to 50% of its effective depth, and D is the duration of the storm in minutes. For small soakaways serving a catchment area of 25 m^2 or less, a design rainfall of 10 mm in 5 minutes may be assumed to give the worst case.

Soakaways for small catchment areas of less than 100 m^2 are usually 3–4 m deep pits filled with granular material having a particle size of around 10 mm to 150 mm, or lined with dry-jointed masonry. NHBC data (see diagram) suggests that, depending on the rate of water infiltration into the ground, smaller pits may be possible. A geotextile covering should be laid over the top, to ensure that no soil is washed down into the soakaway. Those serving larger areas should be designed in accordance with BS EN 752-4 or the latest BRE Digest 365 'Soakaway design'. Generally, these will be lined pits, trenches or perforated precast concrete ring units (to a similar depth as the smaller versions depending on soil characteristics). Loadbearing plastic honeycomb units are also produced to serve the same function. Very deep soakaways may use an access shaft to connect the lower perforated chamber to ground level.

Whichever design is chosen, the soakaway should accommodate an above-ground silt trap and be designed for a return period of once in ten years; that is, they need to be designed to suit the most severe decadal weather event, but taking into account storms of differing durations to determine the one which gives the largest storage volume.

For domestic situations, a pre-inspection by the local authority may not be necessary. However, before backfilling, a Building Control/Approved Inspector together with a representative of the sewerage undertakers and insurer/warranty body should inspect to ensure that the system is functioning correctly. Special care should be taken to ensure that inlet (and outlet pipes or perforations, depending on soakaway design) are free from obstruction and that there is sufficient protection against vehicular access over the top of the pit. A concrete cover may be needed if the chamber is to be concealed beneath the topsoil. These may be either precast concrete components or formed in situ concrete.

In conclusion, the various regional assemblies point out that soakaways 'should be evaluated both with respect to the potential for providing new habitats for protected species (such as water voles and great crested newts) and with respect to the potential for encouraging invasive species'.[1] Make of that what you will.

[1] Highways Agency, Transport Scotland, Welsh Assembly Government, The Department for Regional Development Northern Ireland (2006), 'Design Manual For Roads And Bridges: Design of Soakaways', HA 118/06, Clause 3.2.6, p3/1

References

Bettess, R. (1996) Report R 156 'Infiltration drainage – Manual of good practice', CIRIA.

Highways Agency (2001) HA 89/01 'Design Manual For Roads And Bridges: Environmental design and management. Environmental objectives. Environmental elements', HMSO.

Highways Agency, Transport Scotland, Welsh Assembly Government and The Department for Regional Development Northern Ireland (2006) HA 118/06 'Design Manual For Roads And Bridges: Design of Soakaways', The Stationery Office.

Institute of Hydrology (1994) IH Report 124 'Flood estimation for small catchments', IH.

Kellagher, R. (2000) 'The Wallingford Procedure for Europe: Best Practice Guide to Urban Drainage Modelling', HR Wallingford.

RECOMMENDED READINGS

BRE (2000) Good Practice Guide (GBG) 'Disposing of rainwater', BRE.

BRE (2007) Digest 365 (revised, March 2007) 'Soakaway design', BRE.

Department for Communities and Local Government (2002) 'Building Regulations Approved Document H – Drainage and Waste disposal', NBS.

National House Builders Council (2007) 'Part 5: Substructure and ground floors', NHBC.

35: Wastewater Drainage The real Sanitary Clause

'Is your home polluting the environment?' Or even more accusatorially, 'Are YOU polluting London's rivers and streams?' Thus runs the Environment Agency's Lord Kitchener-esque campaign against foul water contamination of surface water systems. So to avoid being blamed, here we explain the correct method of connecting domestic gravity-fed wastewater drainage.

You have to wonder what some of the European Standards are actually for. For example, BS EN 12056 'Gravity drainage systems inside buildings' is a suite of five parts, but while Part 2 deals usefully with the actual layout and calculations of sanitary pipework, Parts 1 and 3 contain too little information to merit separate documents (nothing that couldn't have been added to Part 2). Part 4 on 'Wastewater lifting plants' is limited in scope, while Part 5 on testing, admittedly, is reasonably handy. However, all of them replicate the glossary and descriptions, and most of them are so generalised as to be verging on trite ('thermal movement shall be considered', and 'noise shall be taken into consideration', for instance). The reason for the vagueness of this European Standard is that national and local regulations are expected to fill in the substantial issues – Building Regulations (England and Wales) and Scottish Technical Standards apply as outlined in the National Appendices. Where national and local regulations aren't referenced or don't exist (after all, the national standards from only 11 out of a total of 25 EU countries are cited) BS EN 12056 hints at a target level of basic provision.

Where national and local regulations aren't referenced or don't exist, BS EN 12056 suggests a target level of basic provision.

"

Scottish Technical Handbook (Section 3: Domestic: Wastewater Drainage), amended in 2007, contains the most contemporary information.

That baseline should ensure that every wastewater drainage system serving a building is 'designed and constructed in such a way as to ensure the removal of wastewater from that building without threatening the health and safety of the people in and around the building.'

According to Thames Water, plumbing mistakes have resulted in waste appliances being connected to the surface water system in around 1 in 10 houses in London (and as many as 1 in 3 houses in some districts). This is eerily reminiscent of the 19th century when Dr John Snow discovered that the cause of cholera – a disease killing more than 10,000 Londoners between 1853 and 1854 – was cross-contamination of the ground water supply with wastewater. Mercifully, we don't draw water from wells any more, but if the figures are to be believed, there is an accident waiting to happen. This Shortcut looks at above-ground wastewater connections for standard domestic arrangements, although Approved Document H (AD H) does stipulate slightly different requirements for domestic buildings over three storeys.

The National House Building Council (NHBC) requires that the installation of internal soil and waste systems be in accordance with AD H. Most of the key information on wastewater above-ground drainage, taken from AD H, is conveyed in the drawing opposite. Once again, the Scottish Technical Handbook (Section 3: Domestic: Wastewater Drainage), amended in 2007, contains the most contemporary information written in a straightforward format. Ironically it relates design and maintenance issues back to BS EN 12056-2. However, there is less core 'technical' information in the Scottish standards and so here we outline the essential requirements of AD H.

PRIMARY VENTILATED STACK SYSTEM

This was previously known as the single stack system. The discharge stack and branch pipes are sized to avoid the need for separate ventilating pipes (see box on page 148). In dwellings especially, this system saves space and costs but must be designed within the limitations of unventilated pipework, summarised in BRE Digests 248 and 249.

The distances of sanitary fittings from the discharge stack (shown) can be exceeded if using resealing (or antisiphon) traps and branch pipe air admittance valves, although these will require access for maintenance.

BS EN 12056-2 (National Annex ND 3.1) allows external pipework on buildings up to three storeys, although internal discharge stacks and branches are preferred. To minimise noise transmission, NHBC Standards (8.1 S8[c]) recommend 25 mm of insulation and a duct casing of at least 15 kg/m² around soil pipes passing through a bedroom or living room. However, locating air admittance valves in sealed boxing will inevitably affect the performance of the valve and BS 8313 clause 12 states that ducts for discharge branches and stacks should be ventilated (except for ducts with a cross-sectional area of less than 0.05 m² in houses, offices and shops, other than food shops).

Where possible, it is good practice to avoid connecting ground floor sanitary fittings and appliances into discharge stacks. This is to ensure that a blockage at the foot of the stack will not overflow through sanitary fittings until it reaches the first floor, by which time the weight of water may have cleared the blockage. Where the system is prone to flood risk, anti-flood devices must be installed along the run of the pipework to the main drain/sewer. To minimise maintenance demands, it is advisable that only those sanitary appliances that are prone to flood risk pass through the anti-flood device.

Offsets in the 'wet' portion of the stack should be avoided wherever possible. In lightly loaded stacks up to three storeys high, they are acceptable without separate ventilating pipes. In other cases, offsets may cause severe pressure fluctuations and require venting above and below. To reduce pressure differentials and the likelihood of blockage, large radius (minimum 200 mm) bends are necessary, and gentle sweeping 'curves' using 45° fittings are preferred.

AD H requires reasonable access to all pipework for repair and maintenance. Discharge branches may be subject to regular sediment blockage when connected to urinals, pipework receiving food or fats, or spray tap basins (prone to soap-gel build-up, especially in soft water areas). Therefore, such branches should be as short as possible, ideally very close to a stack or drain, without bends or gullies, and less than 3 m long.

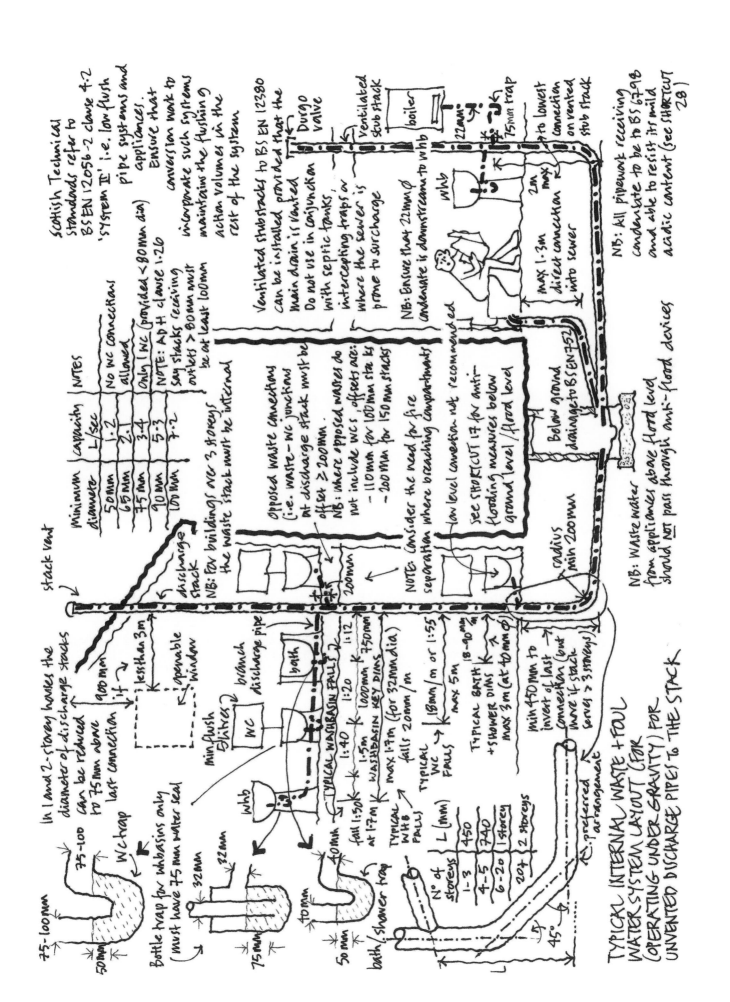

TYPICAL INTERNAL WASTE + FOUL WATER SYSTEM LAYOUT (FOR OPERATING UNDER GRAVITY) FOR UNVENTED DISCHARGE PIPES TO THE STACK

Ventilated branch system This was previously known as the ventilated system or modified single stack system. The stack has a separate ventilating stack, and each branch has a separate full length ventilating pipe connected to it. See BS EN 12056-2, figures 3 and 5, and National Annex NC.3, figures NC.5 and NC.6 (Note: In these documents, these figures are wrongly captioned). This system is used in more onerous discharge conditions, e.g. high-rise buildings with many appliances widely spread. A modified version is possible whereby ventilating pipes are provided only in locations where necessary to protect trap seals, e.g. on long isolated branches.

Secondary ventilated stack system
This was previously known as the ventilated stack system where the stack has a separate ventilating pipe, interconnected at every storey (see BS EN 12056-2, figure 3, and National Annex NC.3, figure NC.4). It is suitable for high-rise buildings and where there are many appliances closely grouped, e.g. back-to-back runs of many WCs, or where below-ground drainage is subject to surcharging.

Gradients should generally be as steep as possible, although (distasteful as it is to mention these things) if they are too steep, solid material can sometimes become stranded. And to avoid self-siphonage, basins with a plug should have waste pipe gradients selected from the graph in AD H Diagram 3b (synopsised on the drawing).

Showers are often installed with the trap and waste pipe at a shallow gradient below the shower tray where access for cleaning is almost impossible. Showers have a relatively low flow rate and can easily become blocked. BS EN 12056-2 (National Annex ND 2.1) states that, for ease of maintenance, a trap may be positioned up to 750 mm from the waste outlet. Alternatively, an access panel or use of a lift-out waste outlet can be provided.

Maintenance operatives should use the correct cleaning agents. Bleach, for example, is problematic for copper pipes, whilst sulphuric and hydrochloric acids erode enamel, and over-use of sodium hydroxide can damage PVC. Eco-pundit, Leo Hickman, writing in The Guardian newspaper says that a weekly tablespoon of bicarbonate of soda followed by a cup full of vinegar down a plughole can help to keep things clear as can a regular flush with boiling water. For old metal pipes whose connections are not designed to expand with heat, this is not recommended.

Coming back to the Environment Agency's concerns, foul drainage must not be connected into rainwater drainage, but where combined drainage exists, it is sometimes a simple and economic solution to connect rainwater from small areas of roof into a foul drainage stack. Before this course of action is considered, the drainage and Building Control authorities must be consulted to ensure that the sanitary pipework and below-ground drainage will not be overloaded during rainstorms.

This Shortcut looks at unventilated systems, the slightly confusing term that actually means those systems that connect to a ventilated stack. Ventilated systems, on the other hand, have secondary pipework to provide additional ventilation in order to limit pressure fluctuations within the system.

References

BS EN 1057 (2006) 'Copper and copper alloys – Seamless, round copper tubes for water and gas in sanitary and heating applications', BSI.

BS EN 1057 (1996) 'Copper and copper alloys – Seamless, round copper tubes for water and gas in sanitary and heating applications', BSI (no longer current but cited in the Building Regulations [England and Wales]).

BS EN 1329-1 (2000) 'Plastics piping systems for soil and waste discharge (low and high temperature) within the building structure – unplasticized poly(vinyl chloride) (PVC-u). Specifications for pipes, fittings and the system', BSI.

BS EN 12056-1 (2000) 'Gravity drainage systems inside buildings. General and performance requirements', BSI.

BS EN 12380 (2002) 'Air admittance valves for drainage systems – Requirements, test methods and evaluation of conformity', BSI.

RECOMMENDED READINGS
BS EN 274-1 (2002) 'Waste fittings for sanitary appliances – Part 1: Requirements' Requirements (AMD Corrigendum 14959), BSI.

BS EN 12056-2 (2000) 'Gravity drainage systems inside buildings. Sanitary pipework, layout and calculation', BSI.

BS EN 12056-4 (2000) 'Gravity drainage systems inside buildings. Wastewater lifting plants – layout and calculation', BSI.

BS EN 12056-5 (2000) 'Gravity drainage systems inside buildings. Installation and testing, instructions for operation, maintenance and use', BSI.

BRE (1981) Digest 248 'Sanitary pipework. Part 1: Design basis', BRE.

BRE (1981) Digest 249 'Sanitary pipework. Part 2: Design of pipework', BRE.

Office of the Deputy Prime Minister (2002) 'Approved Document H: Drainage and waste disposal', NBS.

Scottish Building Standards Agency (2007) 'Scottish Technical Handbook: Domestic 3: Environment', TSO.

The Institute of Plumbing (2002) 'Plumbing Engineering Services Design Guide', The Institute of Plumbing.

36: Below Ground Drainage
Wastewater disposal

BS EN 752 says that 'drains, sewers and other components shall... minimise the use of energy over the life of the system'. In general, the self-cleansing of drains and sewers can be achieved with a pipe diameter (D) of less than 300 mm and with a daily minimum velocity of 0.7 m/s or with a gradient of at least 1:D.

The Building Regulations (England and Wales) state that: 'Foul water means "wastewater" which comprises or includes:

'a) waste from a sanitary convenience, bidet or appliance used for washing receptacles for foul waste; or

'b) water which has been used for food preparation, cooking or washing.'

Similarly, in this Shortcut, the term 'wastewater' includes foul water.

For those who may wish to dip their toes into wastewater, Ofwat, the government-appointed regulator of water and sewerage providers in England and Wales completed the second part of its assessment of the water treatment business. In it, it explored the 'contestability of water and sewerage markets' and the need for 'competition in sewage and sludge treatment' in order to drive investment programmes and hence 'innovation'. Primarily, this innovation relates to coordinated management plans for the network to encourage a reduction in the 3 per cent of total UK CO_2 emissions currently attributable to the provision of water and wastewater services.

> Public sewers in the UK are the responsibility of the ten water and sewerage companies (WasCs); these are public limited companies that came into existence in 1989. All sewers that drain to a public sewer are 'private sewers.'

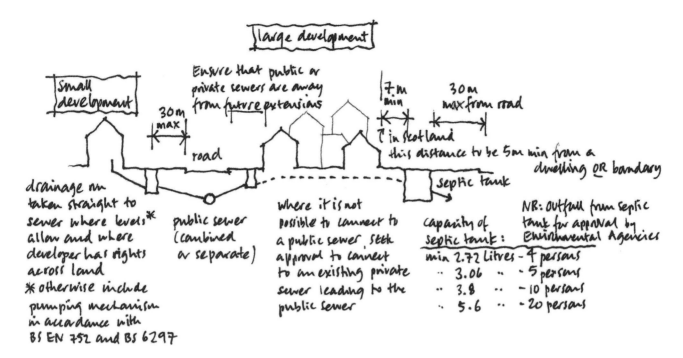

large development

small development

Ensure that public or private sewers are away from future extensions

30m max

road

7m min

30m max from road

* in scotland this distance to be 5m min from a dwelling OR boundary

septic tank

drainage on taken straight to sewer where levels* allow and where developer has rights across land
* otherwise include pumping mechanism in accordance with BS EN 752 and BS 6297

public sewer (combined or separate)

where it is not possible to connect to a public sewer seek approval to connect to an existing private sewer leading to the public sewer

Capacity of septic tank:
min 2.72 Litres - 4 persons
" 3.06 " - 5 persons
" 3.8 " - 10 persons
" 5.6 " - 20 persons

NB: Outfall from septic tank for approval by Environmental Agencies

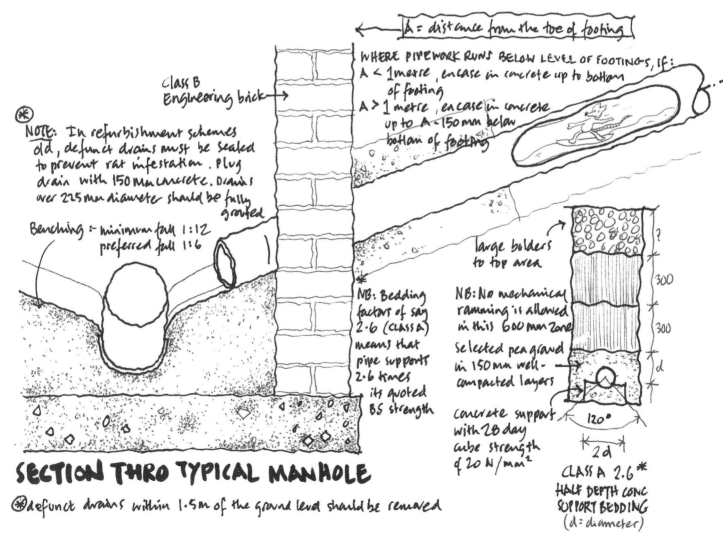

A = distance from the toe of footing

WHERE PIPEWORK RUNS BELOW LEVEL OF FOOTINGS, if:
A < 1 metre, encase in concrete up to bottom of footing
A > 1 metre, encase in concrete up to A-150mm below bottom of footing

Class B Engineering brick →

*NOTE: In refurbishment schemes old, defunct drains must be sealed to prevent rat infestation. Plug drain with 150mm concrete. Drains over 225mm diameter should be fully grouted

Benching := minimum fall 1:12
preferred fall 1:6

NB: Bedding factors of say 2.6 (CLASS A) means that pipe supports 2.6 times its quoted BS strength

large boulders to top area

NB: No mechanical ramming is allowed in this 600mm zone
Selected pea gravel in 150mm well-compacted layers
concrete support with 28 day cube strength of 20 N/mm²

300
300
d

120°
2d

CLASS A 2.6*
HALF DEPTH CONC
SUPPORT BEDDING
(d = diameter)

SECTION THRO TYPICAL MANHOLE

⊛defunct drains within 1.5m of the ground level should be removed

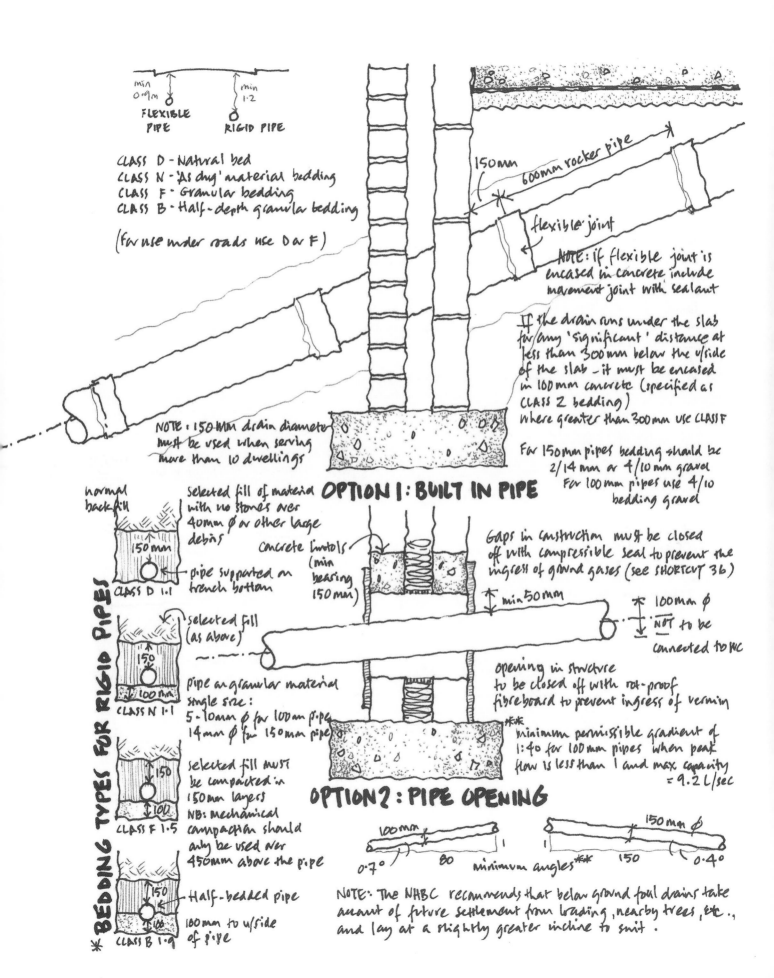

min 0·9m
FLEXIBLE PIPE

min 1·2
RIGID PIPE

CLASS D - Natural bed
CLASS N - 'As dug' material bedding
CLASS F - Granular bedding
CLASS B - Half-depth granular bedding

(For use under roads use D or F)

NOTE: 150mm drain diameter must be used when serving more than 10 dwellings

OPTION 1: BUILT IN PIPE

150mm
600mm rocker pipe

flexible joint

NOTE: if flexible joint is encased in concrete include movement joint with sealant

If the drain runs under the slab for any 'significant' distance at less than 300mm below the u/side of the slab - it must be encased in 100mm concrete (specified as CLASS Z bedding)
Where greater than 300mm use CLASS F

For 150mm pipes bedding should be 2/14mm or 4/10mm gravel
For 100mm pipes use 4/10 bedding gravel

Gaps in construction must be closed off with compressible seal to prevent the ingress of ground gases (see SHORTCUT 36)

min 50mm

100mm ∅
NOT to be connected to WC

opening in structure to be closed off with rot-proof fibreboard to prevent ingress of vermin

** minimum permissible gradient of 1:40 for 100mm pipes when peak flow is less than 1 and max capacity = 9.2 L/sec

Concrete lintols (min bearing 150mm)

OPTION 2: PIPE OPENING

100mm

0.70   80   minimum angles******   150

150mm ∅

0.40

BEDDING TYPES FOR RIGID PIPES

normal backfill
Selected fill of material with no stones over 40mm ∅ or other large debris

150mm
CLASS D 1.1
pipe supported on trench bottom

selected fill (as above)
150
100mm
CLASS N 1.1
pipe on granular material single size:
5-10mm ∅ for 100mm pipe
14mm ∅ for 150mm pipe

150
100
CLASS F 1.5
selected fill must be compacted in 150mm layers
NB: mechanical compaction should only be used over 450mm above the pipe

150
100
* CLASS B 1.9
Half-bedded pipe
100mm to u/side of pipe

NOTE: The NHBC recommends that below ground foul drains take account of future settlement from loading, nearby trees, etc., and lay at a slightly greater incline to suit.

Drains that lie outside the curtilage of a property and connect either to a private sewer that drains to a public sewer, or directly to a public sewer, are known as 'lateral drains' (laterals) and until 2011 these are the responsibility of the individual property owner they serve.

The European Standard BS EN 752 (2008), 'Drain and sewer systems outside buildings', is a framework document to assist in the design, construction, maintenance and operation of wastewater systems. It is supported by a range of other regulations, legislative documents and product standards but has been updated to take account of the forthcoming implementation of the Water Framework Directive, which advocates greater water efficiency. For a 172-page (11.5 MB downloadable file), BS EN 752 says remarkably little, but ties in with the 'Code for Sustainable Homes' (especially Chapters 2 and 4) and is guaranteed to inform water conservation measures which will inevitably be incorporated in the next round of changes to the Building Regulations.

Sewers and drains contain either waste or surface water, and are either public or privately owned. Public sewers in the UK are the responsibility of the ten water and sewerage companies (WaSCs), which are public limited companies that came into existence following the privatisation of the water industry in 1989. All other sewers that drain to a public sewer are 'private sewers' and are the shared responsibility of the owners of the properties they serve. Usually only a small extent of the total length of a private sewer will actually lie in a property owner's own curtilage.

Drains that lie outside the curtilage of a property and connect either to a private sewer that drains to a public sewer, or directly to a public sewer, are known as 'lateral drains' (laterals) and these are the responsibility of the individual property owner they serve. Laterals may lie under private or public land, including highways.

But by 2011, 200,000 km of existing privately owned sewers and lateral drains in England will be transferred into the ownership of these WaSCs. The cost of transfer will mean a rise in household water bills.

In order to calculate the actual functional requirements of a given drainage system, individual appliances which have relatively high, but intermittent flow rates should be assessed by adopting a figure for the peak rate of flow derived from the number and type of appliances connected. BS EN 12056-2 provides the calculations for the rates of flow that these create in the drains. These calculations – or the basic versions included in AD H – should only be used in the design of downstream drain systems. It should be noted that these flow rates can be influenced by national or local regulations and so it is often advisable to contact the relevant authority. For domestic wastewater sewers, flow rates can be based on the rate of flow per head for a given population. Where such data is not available (say, for new developments yet to be occupied) flow rates should be based on the population or the type and number of dwellings set down in the planning criteria.

References

BS EN 1295 (1997) *'Structural Design of Buried Pipelines Under Various Conditions of Loading – Part 1: General Requirements' (incorporating corrigenda May 2006 and July 2008)*, BSI.

BS EN 1610 (1998) *'Construction and Testing of Drains and Sewers'*, BSI.

BS EN 12889 (2000) *'Trenchless Construction and Testing of Drains and Sewers'*, BSI.

BS EN 14654 (2005) *'Management and Control of Cleaning Operations in Drains and Sewers – Part 1: Sewer Cleaning'*, BSI.

HMSO (1994) *'The Urban Waste Water Treatment (England and Wales) Regulations 1994'*, TSO.

Ofwat (2008) *'Ofwat's Review of Competition in the Water and Sewerage Industries: Part II'*, www.ofwat.gov.uk

RECOMMENDED READINGS
BS 476 (1997) *'General Requirements for Components Used in Discharge Pipes, Drains and Sewers for Gravity Systems'*, BSI.

BS EN 752 (2008) *'Drainage and Sewer Systems Outside Buildings'*, BSI.

BS EN 13508 (2003) *'Condition of Drain and Sewer Systems Outside Buildings – Part 2: Visual inspection coding system' (AMD Corrigendum 17163)*, BSI.

Building Regulations (England & Wales) Approved Document H: *'Drainage and Waste Disposal'*, NBS.

Department of the Environment (1990) *Technical Booklet N1: 'Drainage'*, DFPNI.

Hall, F. & Greeno, R. (2007) *'Building Services Handbook'*, 4th edn, Elsevier Butterworth-Heinemann.

NHBC (2007) *'NHBC Standards: Part 5 – Substructure and Ground Floors'*, National House Building Council.

Scottish Building Standards (2007) *Scottish Technical Handbook 3, 'Environment'*, TSO.

Part 5
CPD

37: Testing, Testing
Continuing Professional Development

The General Medical Council requires that doctors complete at least 50 hours of CPD every year and, under the New Practitioners' Programme, barristers must complete 45 hours per annum. While architects have to endure 35 hours, full-time solicitors need only 16. Given that we all need to stay in touch with ever-changing rules, regulations and practices, this Shortcut tests you out.

The RIBA requires at least 17.5 hours 'official' CPD training from a syllabus of approved core curriculum subjects and a minimum additional 2 hours of compulsory health and safety training. The remaining minimum of 15.5 hours can be attributed to maintaining and enhancing a general awareness of professional care and judgement. The RIAS has similar requirements. The evidence of compliance should be recorded (and future CPD needs planned out) on the RIBA's record sheet and professional development plan respectively (online where possible). At the end of every year, the RIBA monitors a random sample and delivers a few rapped knuckles.

For many of us, real study is a thing of the past. Professional qualifications are a dim and distant memory. Lots of us have no real idea what PEDR[1] stands for and Donoghue vs Stevenson probably sounds like a Scottish light-heavyweight contest from the 70s. Conversely, for those sitting RIBA Part III, you may well be versed in the minutiae of clause 7.1[2] of JCT Intermediate Form of Contract, 1998 edition and will be able to recite 'The Plan of Work', but do you know where the kettle is? So, for young and old, musing over past papers provides a refreshing opportunity for rusty architects and unworldly students to see how sharp they really are.

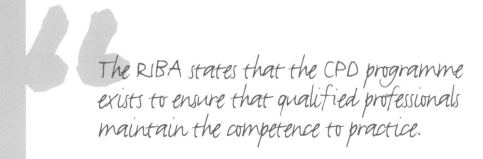

The RIBA states that the CPD programme exists to ensure that qualified professionals maintain the competence to practice.

In the 'Part 3 Handbook', Stephen Brookhouse categorises the range of final exam questions into the following headings:

1. knowledge-based
2. problem/scenario-based
3. issue-based

In the spirit of non-dumbed-down yet simplified articles, this Shortcut presents some past papers and general questions that, if completed, should account for about 2 hours of CPD.

QUESTION ONE
(Knowledge-based)

Explain the following terms (with reference to a particular contract)

- practical completion
- final certificate
- liquidated and ascertained damages
- partial possession
- sectional completion
- handover
- defects liability period
- CDM coordinator
- tort
- liability

QUESTION TWO
(Problem/scenario-based)

THE SCENARIO

You are the sole practitioner in an architectural practice in Chigley, a small market town outside Trumpton. You are mainly involved in residential developments – new-build housing and refurbishments. Your clients tend to be shop owners, windmill operators and small developers, and you have decided not to get involved in speculative development or financial arrangements to fund projects, preferring the relative simplicity of maintaining your 'traditional role' as architect.

However, you have the opportunity, through a contact at Winkstead Hall, to design an extension to the existing police station; to include new cells, changing facilities, WCs, reception, meeting rooms, plant, stores and kitchens. The works will involve reorganising the majority of the existing functions within the original building, extending it from 425 m^2 to 1150 m^2, and modifying services accordingly. The client's committee assumes the works will cost £1.5 million, although there are cash reserves to cover an additional £100,000 for what it calls 'all eventualities'.

On your first meeting with the client – a committee made up of five appointed individuals delegated from the local constabulary and lay members – you are informed that the police station and car park will have to remain in operation throughout the works and the works must be finished in 3 months. The committee members describe the general extent of the works.

They also insist that you tell them there and then your level of fees so that they can budget for the works and get things moving quickly.

- What immediate issues should you raise at this meeting?
- What should be your first course of action?

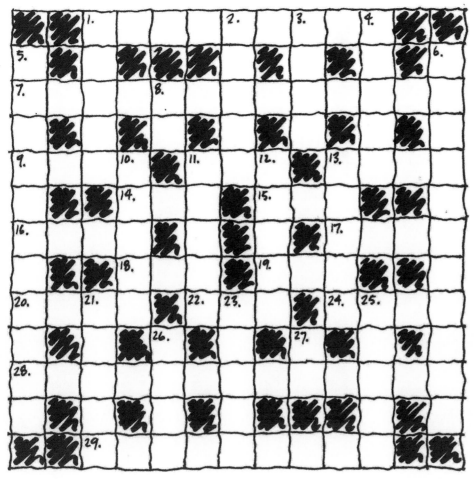

CRYPTIC CPD CROSSWORD*

Across:

1. A confused milieu ran into the complete light fitting (9)
7. Testing times in a fake firmament (10, 3)
9. The likelihood of a baby biscuit replacing you with me (4)
11. Approved Document L1's notional guidance value in kg/m²/year (3)
13. Joel Garreau's cutting city (4)
14. Jo Coenen's cultural facility in Rotterdam (3)
15. It sounds like the smell of sweat on an 2012 athletics' organiser (3)
16. A house in Spain (4)
17. Carbon-based chemicals that easily vaporise at room temperature (4)
18. PVC blended with ethylene vinyl acetate (3)
19. Magazine that confuses the new engineering contracts (3)
20. Bathroom, roof, Scrabble, mosaic, cutter or nibbler (4)
22. H&S benchmarkers working with the CIC (3)
24. Letters for the critical evaluation of the threat of water inundation (4)
28. Currently averages 2.4 persons per household, 1.64 per car, and 90 per cent per hospital (9, 4)
29. Led Zeppelin's act raises confusion in song title; not heavenward, but leading nowhere just for show (9)

Down:

1. The acanthus of the east inspires Dubai development (5)
2. Hiding place, and small-scale Classical exedra (5)
3. Avery's picture housed in rotunda, showing at Osaka's 1970 Expo (4)
4. Please don't. In the middle of it all, the planners reduced their demands (5)
5. A falsehood about construction (11)
6. Virtual places (11)
8. Neither higher nor broader, but shortened education (2)
10. Scoundrel in the middle of the church (5)
11. Hugh's mis-spelled shades (5)
12. Not arched or square, but see through (5)
13. From whence Vitruvius' stillicidium arise (5)
21. Yale's chubby operators lever a way in to key positions (5)
23. Stirring renin into blockwork leaf. Probably not the outer brickwork leaf (5)
25. Set up a house and stitch up an architect (5)
26. The UK's polymeric waterproofing authority (4)
27. A house in Wales, also with pots and leaves (2)

*Allow a maximum of 35 hours to complete
For crossword answers see page 161.

On returning to your office, you receive a phone call from Police Constable McGarry (Number 452), one of the committee members not delegated to the project meeting. He says that he disagrees with the committee's assessment of the amount of work needed to be done. The WCs are 'more than adequate' and he says that the kitchen is a separate franchise so any works to be done on this part of the building will not be funded by the force. He also tells you, in passing, that the windows were installed 'only 10 years ago' and should not be a problem but the carpet definitely needs to be replaced, and his brother is a supplier and can get any length of 'reasonably hard-wearing' carpet for 20 per cent less than the normal retail price.

■ What should be your course of action?

Twenty-four hours later, the committee rings to ask for some general inception sketches to be done and that you have to make a presentation to the local policeman's ball on the weekend, in two days' time. This open meeting will comprise about 75 people and a jazz band and will take place at 7pm. You are asked to bring along the design team, to present some slides during the interval before the comedian, to confirm the work involved and the contract period, and to state the cost of the works. PC McGarry verbally agrees on the phone that they will accept an estimate at this stage and will not hold you to it.

■ What should be your course of action?

■ What contract period would you recommend?

■ How do you comprise and advise your design team?

■ What will your fee be at this stage, and overall?

(continued overleaf)

The six areas of RIBA's CPD Core Curriculum are: health and safety, professional context, practice management, managing projects, construction skills and personal skills.

QUESTION THREE:
(Issue-based)

With reference to the previous scenario:

- Describe what you would do in any dispute with the client that gives rise to a complaint to the ARB against you personally.

- If the client asks for a collateral warranty from you specifically to cover the flooring work, what should you do?

- Given that the site abuts the main road at the side and a school building to the rear, what actions should you consider under the Party Wall etc. Act 1996?

- What aspects of the RRO[3] apply?

[1] PEDR is the RIBA Professional Experience and Development Record. It is the online professional experience record for architectural students on the year out and post Part 2 working towards their Examination in Professional Practice and Management (Part 3). More information and samples can be found on www.pedr.co.uk/

[2] A valid notice served on a party to a contract 'shall be in writing and given by actual delivery, special delivery or recorded delivery'. The case of Construction Partnership v Leek Developments [2006] CILL 2357 shows that the issuance/receipt of faxes is a valid delivery mechanism; and Bermuth Lines v High Seas Shipping (2005) confirms that it also applies to email correspondence.

[3] Regulatory Reform (Fire Safety) Order introduced in October 1996. (See Shortcut: Book 2)

References

Cornes, D. and Winward, R. (2002) *'Winward Fearon on Collateral Warranties'*, Blackwell Science.

Green, R. (2001) *'Architect's Guide to Running a Job'*, 6th edn, Architectural Press.

Hyett, P. (2001) *'In Practice'*, Emap Construct.

Dalziel, B and Ostime, N (2008) *'Architect's Job Book'*, 8th edn, RIBA Publishing.

Lupton, S. (2001) *'Architect's Handbook of Practice Management'*, 7th edn, RIBA Publishing.

Nicholson, P. (2002) *'Architects' Guide to Fee Bidding'*, Spon Press.

RIBA, 'Guide to the RIBA Agreements 2007', RIBA Publishing.

Williams, A. (2009) *'Shortcuts: Book 2 – Sustainability and Practice'*, RIBA Publishing.

RECOMMENDED READINGS

Association of Consultant Architects, (2008). 'Standard Form for the Agreement for th Appointment of an Architect, ACA SFA/08', ACA.

Brookhouse, S. (2007) *'Part 3 Handbook'*, RIBA Publishing.

Chappell, D. (2003) *'Legal and Contractual Procedures for Architects'*, Butterworth-Heinemann.

Chappell, D. (2003) *'Standard Letters in Architectural Practice'*, 3rd edn, Blackwell Publishing.

Phillips, R. (2000) *'The Architect's Plan of Work'*, RIBA Publishing.

The RIBA Outline Plan of Work 2007 and details of the Core Curriculum for CPD are available on: www.riba.org

SHORTCUTS: BOOK 1
STRUCTURE AND FABRIC

INDEX

Handwritten notes

Scottish Technical standards refer to BS EN 12056-2 clause 4.2 'System II' i.e. low flush pipe systems and appliances. Ensure that conversion work to incorporate such systems maintains the flushing action volumes in the rest of the system

75-100mm

In 1 and 2-storey houses the diameter of discharge stacks

stack vent

Minimum diameter	Capacity L/sec	NOTES
50mm	1.2	No WC connections allowed
65mm	2.1	
75mm	3.4	Only 1 WC (provided < 80mm dia)
90mm	5.3	NOTE: ADH clause 1.26
100mm	9.2	Say stacks receiving outlets > 80mm must be at least 100mm

75-100 can be reduced to 75mm above last connection if

900mm

WC trap

Bottle trap for washbasins only (must have 75mm water seal

discharge stack

NB: For buildings over 3 storeys the waste stack must be internal

if less than 3m

if openable window

minflush 8 litres

WC

branch discharge pipe

32mm

32mm

whb

50mm

75mm

Ventilated stub stacks to BS EN 12380

Solution for crossword on page 157

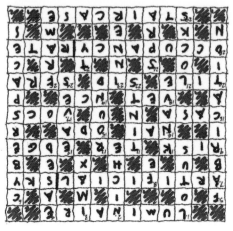